I0112579

Ecology and Diversity

Biology Lab Manual

Second Edition

Chemeketa Community College Biology Faculty

Edited by Jennifer Johns

Chemeketa Press | Salem, Oregon

Ecology and Diversity: Biology Lab Manual
© 2024, 2025 by Chemeketa Community College

ISBN-13: 978-1-955499-63-7

All rights reserved. Edition 1.0 2024. Edition 2.0 2025.

No part of this book may be reproduced or transmitted in any form or by any means, electronic or mechanical, including photocopying, recording, or by any information storage and retrieval system, without permission in writing from the publisher.

Chemeketa Press
Chemeketa Community College
4000 Lancaster Dr NE
Salem, Oregon 97305
collegepress@chemeketa.edu
chemeketapress.org

Cover design by Ronald Cox
Interior design by Ronald Cox and Abbey Gaterud

References to website URLs were accurate at the time of writing. Neither the author nor Chemeketa Press is responsible for URLs that have changed or expired since the manuscript was prepared.

Printed in the United States of America.

Land Acknowledgment
Chemeketa Press is located on the land of the Kalapuya, who today are represented by the Confederated Tribes of the Grand Ronde and the Confederated Tribes of the Siletz Indians, whose relationship with this land continues to this day. We offer gratitude for the land itself, for those who have stewarded it for generations, and for the opportunity to study, learn, work, and be in community on this land. We acknowledge that our College's history, like many others, is fundamentally tied to the first colonial developments in the Willamette Valley in Oregon. Finally, we respectfully acknowledge and honor past, present, and future Indigenous students of Chemeketa Community College.

Contents*

---

* Note: The labs for this class occur in a different order depending on the term due to the timing of the field trip(s). Please consult the lab schedule in your course syllabus to determine which lab you need to prepare for.

Lab and Safety Regulations

A. Personal Behaviors

1. Eating and drinking are prohibited in the labs.
2. Wash your hands before leaving the lab.
3. Wash the lab counter & related work areas with disinfectant before and after the lab.
4. Stow your personal items in the cubbies or under your station.
5. Keep the lab counters and work areas uncluttered.
6. Clean-up is your responsibility:
   a. Return cleaned items back to lab kits.
   b. Wash slides and return to the original location (unless told otherwise). Cover slips may be thrown away.
   c. Wash glassware and return to original location (unless told otherwise)
   d. Ensure any lab waste is disposed of in the approved container(s).
   e. Ensure sinks and counters are as clean as when you arrived.
7. Clothing must be appropriate for the lab to be performed.
8. Read the lab in advance so you're aware of proper safety protocols.
9. If you have questions about safety, ask before you do.

B. Safety Protocols

1. Know where the lab safety equipment is located–fire extinguisher, first aid, eye wash, etc.
2. Wear personal safety equipment (goggles, gloves, aprons) as indicated in the lab instructions or by your instructor.
3. Handle all chemicals and biologicals, including stains, below eye level.
4. Do not assume something can just go down the sink or in the trash. Dispose of wastes in appropriate waste containers.
5. If there is a chemical spill:
   a. Get your instructor.
   b. If the spill is on the floor or counter, keep away from it.
   c. If the spill is on you or your clothes, rinse the area immediately with running water.
   d. If you splash chemicals in your eyes, go immediately to the eye wash station, turn on cold water, remove the red caps and lean down so that the water bubbles into your eyes.
6. In case of injury:
   a. Get your instructor.
   b. If the instructor is not available, call campus safety (x5023) on the lab's phone.
7. If the injury appears severe, also call 911
8. In case of fire:
   a. If your clothes are on fire, yell "FIRE" and roll on the floor or use a coat or fire blanket to smother it.
   b. If chemicals or lab equipment is on fire, call out "FIRE" and evacuate the room.
9. If the fire alarm goes off, take your essential items and leave the room. Head to the evacuation area as told by your instructor. Stay with your class and wait for instructions. Do not re-enter the building until an all-clear is given.

Environmental Issues Project

A generation ago, when the world was not quite so connected every moment of every day, there was a saying that went: "Think globally, act locally." As this saying suggests, there are global environmental issues that humans have caused that each of us needs to understand, and yet the place where most of us will have the biggest impact is locally, in our own houses, yards, city parks, nature trails, and state parks. And while one might be very concerned about the deforestation of the Brazilian Amazon, understanding that the removal of this enormous tract of forest will change weather patterns around the globe, one could personally make a difference regarding deforestation by planting trees on one's own property—or lacking property, one could volunteer with a local organization, such as Friends of Trees, and plant trees at a local park. Think globally, act locally. This is the challenge we present to you this term for your Environmental Issue Project.

Possible Project Types

The project you choose should be interesting and meaningful to you. You will work with a group of other students, ideally in a group of three or four. There are two main choices for your project: (1) an environmental action project or (2) a community science project. For both types you will research the global extent and impact of the issue and then focus on how the issue plays out in Salem, or the Willamette Valley, or in Oregon. You will then organize and carry out your project to make this problem a little bit better locally.

If you conduct an environmental action project, you will be volunteering at least 3 hours of your time to an organized event or action you design. For example, you might choose to plant trees, remove invasive species, or create fish habitat in streams. If you conduct a community science project, you will be conducting a research study of the issue. For example, you could quantify and compare microplastic pollution in different streams around town, investigate the number and health of street trees in different neighborhoods, or tally the number of native bees observed in a restored habitat. Regardless of the type of project you choose to address your environmental issue, you will have the same set of assignments to complete as outlined in the list below.

Environmental Issues Project Assignments

1. An **article review** of a local environmental problem seen in the Willamette Valley or Oregon is due at the end of Week 1 or beginning of Week 2 to inform a class discussion and create groups.
2. Your **group proposal** for an environmental action project or a community science project is due in Week 2.
3. A **final plan** of what you are doing, when you are doing it, and with whom is due during Week 3.
4. A **group research paper** and an **individual reflection** are due in Week 7.
5. Your **near-final draft of your powerpoint or poster presentation** is due in Week 8 in Fall term (due to the Thanksgiving holiday), and Week 9 in Spring term.
6. **Final presentations** will be made DURING LAB TIME in Week 10
7. **Review of classmates' powerpoints** and an evaluation of your teammates are due at the end of Week 10.

> **Note:** The details of your actual assignments will be provided and explained in greater detail by your instructor. Every student in BI101 must participate in the environmental action or community research project to pass the class. (See below.)

Proof of Participation

You are required to **submit proof of your participation** in your environmental action or community research activity. You will be provided with an Individual Journal Worksheet that includes space for the signature and contact information for the volunteer coordinator for your activity. Talk with your instructor about other methods of providing proof of participation.

Reference Format Requirements for General Biology

This class includes a research project and with research comes references. To avoid charges of plagiarism, it is essential that you learn the proper way to cite the source of information in the text of your paper and list complete references at the end of your paper. Plagiarism will result in a score of zero on your assignment. You may already be plagiarizing without knowing it. Get informed. Visit https://go.chemeketa.edu/whatisplagiarism or scan QR code for a simple explanation of what is and is not plagiarism.

Using References in Science

For the most part, **scientists do not quote directly from papers but summarize the facts or conclusions of a paper in one or a few sentences** that are referenced at the end of a sentence. The source for these summarized facts is then referenced at the end of the sentence using an intext citation that includes the author and date of publication. See below. You must practice summarizing previously published material in your own words. **Points will be taken away from your grade for using quotations rather than summarizing your references.** After summarizing the data, cite the publication to which you are referring.

In Text Citation format: …(Author, Year).

- ❏ Scientific citations include the author and the year of publication but not the page number.
- ❏ Citations are placed in parentheses at the end of a sentence before the period.
- ❏ If one citation covers several sentences, you either place the citation at the end of each sentence or put the citation at the end of your summary (this is the best, most appropriate way to do it).

Your Reference List

Any resources cited in your text must be listed in proper format in the References section of your paper. Your reference list allows the reader to find more information on the topics you cited in the text of your paper and provides verification for your discussion section. Most scientific publications require a format based on, but not exactly the same as APA (American Psychological Association) format. Examples of format required for this class are provided on the next page.

C. Reference List Format

❑ For the format of different types of references, use the following guide.

❑ References should be alphabetized by author's last name. Notice that all formats list author last name first.

❑ Only references actually cited in your paper should be included in the reference list.

Required Reference Format for General Biology

1. **Journal articles** (including those accessed digitally)

    Author(s) (year) Title of article, *Journal name* vol: pages.

    Freas KE, PR Kemp (1983) Some relationships between environmental reliability and seed dormancy in desert annual plants, *Journal of Ecology* 71: 211–217.

2. **Books**

    Authors(s) (year) *Book title*, Publisher, City, State, and country where publisher is located.

    Fenner, M (1985) *Seed ecology*, Chapman and Hall, New York, NY, USA.

3. **Edited books**

    Author(s) (year) Chapter title. In *Book Title* (names of editor(s), eds.), Publisher, City, State, and Country, pages

    Kemp PR (1989) Seed banks and vegetation processes in deserts. In *Ecology of Soil and Seed Banks* (MA Leck, VT Parker, RL Simpson, eds.), Academic Press, New York, NY, USA. 257–282.

4. **Websites**

Author(s)* (year†) Title of Webpage. Retrieved on <date you visited that website> from <complete website URL>.

Backyard Gardener (no date) Seed Germination Database. Retrieved on February 8, 2011 from http://www.backyardgardener.com/tm.html.

5. **Generative AI Websites** (for example, ChatGPT)

Name of the generative AI site. (Date queried). "Copy of the Query/Request you put into the Generative AI site." Generated using Open AI. <website link>

Notice that book and journal titles are always italicized or underlined. Chapter titles and article titles are in normal font and only the first letter of the first word is capitalized.

---

*   Find the author at the very bottom of the page. If no author listed, use the sponsor of the website (check the URL for the name before .com/.org/.gov)

†   Some webpages lack a date of publication. If this is the case, put "no date" in the parentheses. You should ask yourself if this really is a reliable resource if it doesn't give you a date!

1 | The Scientific Method Pre-lab

1. Put the steps of the scientific method in order by placing the appropriate number (1–8) in the blank next to the step.

   ❏ Analyze results_____
   ❏ Conclusion _____
   ❏ Design Experiment _____
   ❏ Hypothesis _____
   ❏ Observation _____
   ❏ Prediction _____
   ❏ Run Experiment_____
   ❏ Question _____

2. Does the scientific method really occur as the linear process like the one shown above? Explain your answer.

1

3. Define the following terms related to experimental design.

   ❑ Controlled variable:

   ❑ Dependent variable:

   ❑ Independent variable:

4. What is the difference between a control group and a controlled variable?

5. Briefly describe an experiment you conducted in middle or high school.

1 The Scientific Method

The procedure used by scientists to generate knowledge is known as the scientific method. Although many textbooks describe this method as a linear process, it is really more of an on-going cycle of observing, asking questions, and proposing explanations, testing those explanations, and analyzing data, as seen in figure 1.1. You can start anywhere in the cycle and go in either direction (clockwise or counterclockwise). The key is to remember that all aspects of the scientific method rely on observable (by eye or with the aid of a tool like a microscope) and measurable data that provide biologically valid explanations.

Figure 1.1. The Scientific Method

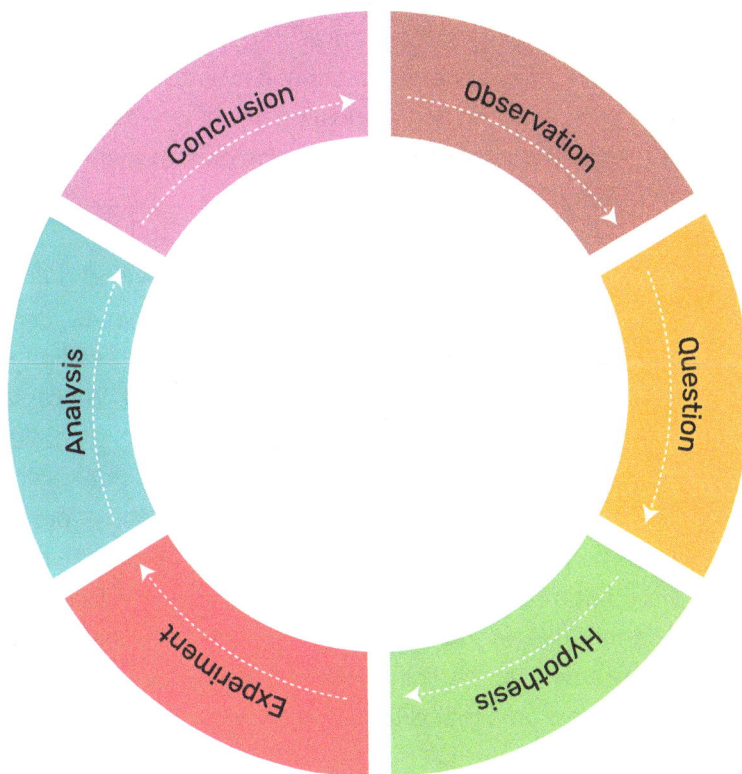

In this lab, we will focus on using the scientific method to develop and run a simple experiment. Throughout the term, you will continue to use this method to observe, ask, propose and test your own ideas as part of group and individual projects.

By the end of this lab, you should be able to:

- ❑ Formulate a hypothesis that meets the conditions of being falsifiable and testable.
- ❑ Differentiate between a hypothesis and a prediction.
- ❑ Design an experiment to test a hypothesis.
- ❑ Indicate the independent, dependent, and controlled variables in an experiment.
- ❑ Design a table and a graph to display experimental data in a meaningful way.
- ❑ Draw a conclusion based on scientific data.
- ❑ Communicate the results of a scientific experiment to peers in a group setting.

Exercise 1.1. Observations, Questions, and Hypotheses

Humans commonly ask questions and propose answers based on what we know and have observed. The scientific method is part of our everyday life. Scientists, however, focus their questions and observations on the understanding of the natural world. Scientists cannot answer all questions. For a scientist to attempt to answer a question, it must be related to a natural and testable phenomenon. While a scientist can test a question about whether genetically modifying a crop plant can reduce the need for pesticides, he or she cannot test a question about whether the act of genetic modification is ethical.

Scientists answer questions about natural phenomenon with explanations based on observations and previous knowledge. These proposed answers are called **hypotheses**. Just as scientists are limited to particular types of questions, they are also limited to specific types of hypotheses. A scientifically valid hypothesis must be testable and falsifiable. **Testable** refers to our ability to develop an experiment to determine whether or not the explanation is valid. The hypothesis *should not predict* what will happen in an experiment, but explain the concepts that make it a possible answer to the question. **Falsifiable** means the experiment must be able to produce results that do not support the explanation. Additionally, the hypothesis must be clearly written in a way that makes what will be measured to test that hypothesis clear. That is, the hypothesis cannot use vague words like "better" or "easier," or even "more." As in: People who are given the new cold medicine will get better sooner than people who take the old medicine. Rather the symptoms that will be measured over what time frame would need to be specified in the hypothesis.

Q1. List and describe the two required criteria of a valid scientific hypothesis.

Q2. Consider each of the following statements. Put a check next to the hypothesis that meets the requirements for being scientifically valid.

❑   I won the lottery because I am lucky.
❑   I received an A on my biology exam because I studied every day.
❑   My keys keep disappearing because there is a ghost in my house.
❑   No human lives forever.

Q3. Explain why the statement "Adding fertilizer will make plants grow faster" is not a valid hypothesis.

Q4. Write a scientific hypothesis that would answer each of the following questions. When you're done, check with a classmate. Did you actually write a hypothesis or a prediction? Sometimes it helps to write a hypothesis in the following format: I hypothesize that…because…

❑   Why are the leaves on my houseplant turning yellow?

❑   Only one sock came out of the dryer. What happened to the other sock?

❑   Why doesn't the light turn on when I flip the switch to on?

Procedure

1. Depending on the term and season, different materials will be provided for you to conduct an experiment with. These may range from small organisms like rollie pollies or worms to household objects to outdoor observations.
2. Read any material or watch any videos provided to your group about this subject.
3. Record observations and questions *you* have about this subject in the space provided.

Observations

Questions

4. Discuss your observations and questions with your group.
5. As a group, choose one question that you are interested in testing.
6. List that question in the space provided below.
7. Propose a hypothesis that answers your question. Your hypothesis does not have to be the same as that of other group members, but every group member must have a proposed and written hypothesis to move on.

Group Question

Your hypothesis
(Make sure it is testable and falsifiable.)

Check with your instructor before moving on to the next exercise. **Initials:** _____

Exercise 1.2. Designing an Experiment

Once a scientist has chosen the best hypothesis, they must test that hypothesis. Scientists use many processes to test hypotheses. You are probably most familiar with the idea of a **controlled experiment**. This type of test involves changing just one variable at a time and observing the outcome. The development of this type of experiment requires the scientist to define a specific set of variables:

❑ **Dependent variable:** This is the factor that will be measured in your experiment. It is considered "dependent" because the measurements "depend" on the treatment.

❑ **Independent variable:** This is the factor that your hypothesis focuses on. It is the thing that changes between the different samples in your experiment. The independent variable is a generalization of your treatments. For example if you treatments are 5g or 10g or 15g of fertilizer (to see which produces the largest weight of tomatoes), the the independent variable is the amount of fertilizer.

❑ **Controlled variables:** These are the factors that must be kept constant for all the samples in your experiment. When you design an experiment, you want to ensure that only your independent variable is affecting your results. All other factors must be kept the same between all the samples. For example, if you are testing the effect of fertilizer on plant growth, you must have all the plants receive the same amount of sunlight.

Many experiments also include a **control group**, which is a sample in your experiment in which the independent variable is never changed. The control provides a baseline for comparison and allows you to determine whether the independent variable has had an impact on your dependent variable. An experiment can include many types of controls. In this class we will generally perform experiments that have a positive control, a negative control or both. A **positive control** is a sample that shows what the expected result should look like. A **negative control** is a sample that shows what "no effect" looks like. For example, in the fertilizer experiement suggested above, the control would be a group of plants that received 0g fertilizer.

Q5. Describe a control group for each of the experiments below and state whether it is a positive or negative control. Additionally, state which variables could or should be controlled or kept constant to ensure a "fair" comparison.

❑ A plant biologist wants to know how a particular plant hormone affects plant cell growth, so she treats different plants with different concentrations of the hormone.

» Control group:

» Controlled variables:

❑ An ecologist is trying to determine the impact of an invasive species on an oak savannah so he removes the species from a specific region of the oak savannah.

 » Control group:

 » Controlled variables:

❑ A molecular biologist uses genetic engineering to introduce a gene that promotes insect resistance from tomatoes into corn. They want to test if the corn is really resistant.

 » Control group:

 » Controlled variables:

Once an experiment is described, it is important to double-check that it actually tests your hypothesis. To do this, scientists make a prediction as to what will happen in the experiment. Remember, a **prediction** is an if-then statement that says "If my hypothesis is true then this is what I should see in this experiment." If you are unable to form a prediction based on your hypothesis and experiment, you probably need to go back to the drawing board.

Q6. What is the difference between a hypothesis and a prediction?

Q7. Make a prediction about one of the experiments above. (*Hint: you will have to come up with a hypothesis first.*)

In addition to an independent variable, dependent variable, control, and controlled variables, scientists must also perform their tests on more than one individual to be certain that the outcome of the experiment is due to the independent variable (and not some weird characteristic of the individual that received the treatment.) Each of the individuals receiving the same treatment is called a **replicate**. The concept of a replicate is most easily exemplified in considering a test of a new drug. Scientists who are trying to conclude if a new drug is effective in treating cancer would never test it on only one person. If you read in the paper that one person was tested and cured and so the FDA has gone ahead and approved the drug for widespread use, you'd think the test is inadequate at best! You'd want the drug to be tested on hundreds, if not thousands, of people before you felt the research adequately demonstrated the effectiveness of the drug. In this scenario, each person in the experiment is a replicate. As you design your experiment below, be sure to incorporate **replicates** for your treatments.

Procedure

1. Complete table 1.1 by describing the control and experimental treatments and indicating the independent, dependent, and controlled variables for your experiment.

Table 1.1.

Independent variable (In general, what are you going to change?)	
Experimental treatment(s) (How will you change the independent variable?)	
Dependent variable (What are you going to measure?)	
Control group (Against what are you going to compare your results?)	
Controlled variables (What do you have to keep the same across treatment groups so that the experiment is "fair?")	

2. Write a step-by-step methods procedure in the box provided that indicates precisely what you are going to do to your treatment group(s), for how long and/or for how many times and describes exactly what and how you will measure the effect.

> **Step-by-step methods**

3. Write a prediction for the outcome of your experiment (based on your hypothesis).

> **Prediction**

You might feel like you are ready to get started on your experiment. You are nearly there. First you must work through the next exercise to create a data sheet to record your data.

Exercise 1.3. Gathering and Displaying Data

Running an experiment is the easy part. Once the data is gathered, scientists spend a great deal of time trying to interpret their results. You may think that the results of an experiment are always straight forward, but that is rarely the case. In this section, we will learn about how to display your data in a meaningful way.

A **table** is a chart based on rows and columns where data, numeric or otherwise is organized by category. Tables generally do not show the trends in a dataset but are convenient for recording data and displaying data that cannot be graphed. In this exercise, you will create a data table to record the results of your experiment.

A **figure** is any type of visual data display that does not involve a table including graphs, photographs, diagrams, and drawings. Figures tend to be a better way to show the trend in your data. This exercise focuses on the use of graphs. When you display your data in a graph, you must also determine what type of graph to use. Many types of graphs exist, but most often, you will need to choose between a bar graph and a line graph.

- ❑ Use a **line graph** when:

  - » Both variables, independent and dependent, are numeric.
  - » The data is continuous (e.g. measurements are taken multiple times from the same sample at specific intervals).
  - » The differences between the data points occur on a small scale (you can adjust the scale on the axis to make the differences more obvious).
  - » You are plotting more than one sample on the same graph.

- ❑ Use a **bar graph** when:

  - » One of your variables is not numeric.
  - » Your data is non-continuous data or category-based.
  - » Your data is continuous data and the change between data points is large.
  - » You want to show categories relative to each other.

- ❑ Use a **pie chart** when:

  - » Your categories of data are part of a whole.
  - » Each category makes up a percent of the total.
  - » Together the categories total 100%.

Notice that there is some overlap between these graph types. The key is to determine *which format best displays the trends you see in the data*. In this exercise you will use a graph to show the trends in the data you collected in your experiment.

Procedure

1. Design a table in the space entitled "table 1.2" to gather the data for your experiment. The data table for every member of your group should look the same. Be sure to follow the directions about what to put in the columns and rows. A template is not provided for you because different groups will have different numbers of columns and rows. Use a ruler if necessary.

    Notes on table format:
    »   Provide a descriptive title for your table. What is your table about?
    »   Different experimental/control treatments are usually listed in the left-hand column.
    »   The first row of the table should list the dependent variable (including units).
    »   If several measurements are made of the same treatment (measuring the temperature every 15 minutes, for example), additional columns are added.

Table 1.2. Table Title _____

    Get your instructor's initials to review your step-by-step methods and data table before moving on. _____

2. Run your experiment and complete the data table as you work.

3. Determine whether to use a line graph or a bar graph to display your data. Use your data to create the graph in figure 1.2. Create a graph below to record the averaged results from your experiment. Be sure to follow the directions about what type of graph would be best for your results. Don't forget to include a very descriptive title and to label your axes.

Notes on graph format:
- » All graphs need a descriptive title. What is the graph about?
- » The independent variable always goes on the X-axis and the dependent variable always goes on the Y-axis.
- » Choose a scale that is appropriate for each axis (your data points should fill your graph rather than being squashed together at one end).
- » The X- and Y-axes should be clearly labeled and units of measurement should be indicated.
- » If a graph has more than one line, the different lines should be clearly distinguishable by color and/or line texture and a key should be provided.
- » Numbers should not be attached to data points.

Figure 1.2. Graph Title _____

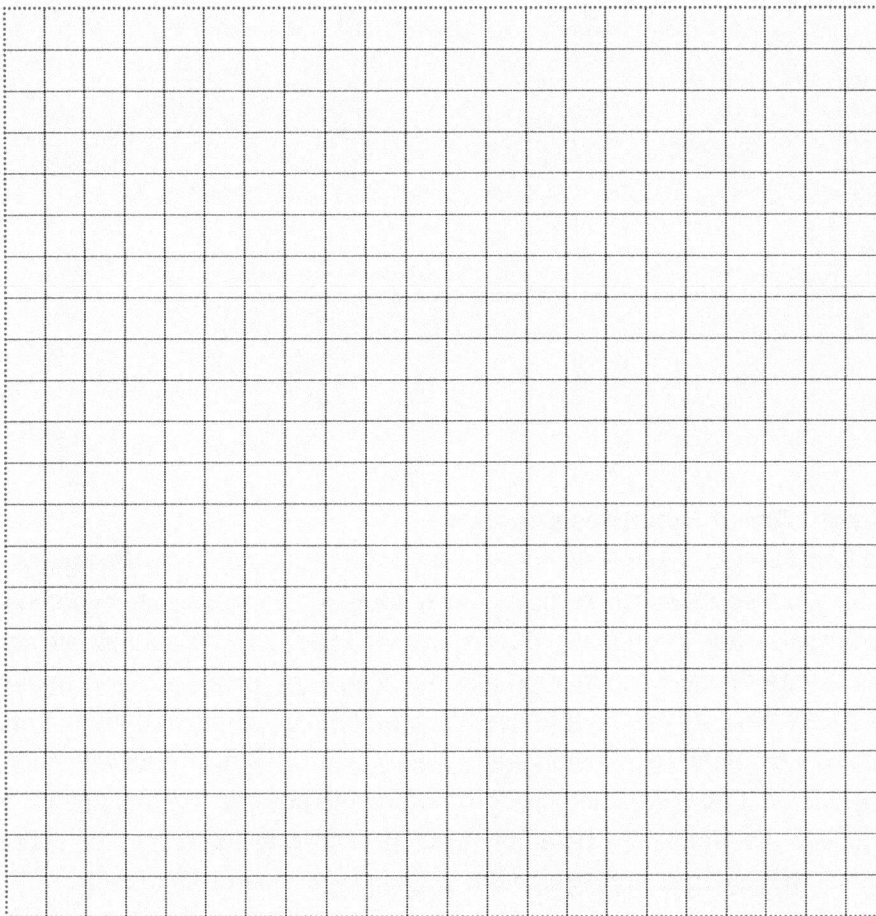

Exercise 1.4. Communicating Your Results

Communication is a key part of being a scientist. Scientists rely on feedback to decide what questions to answer, design hypotheses, design better experiments, interpret their data, and draw conclusions. Scientific communication occurs through a variety of processes including less formal discussions at lab meetings or at the bar during a conference and more formal presentations at meetings and the publication of scientific papers. In this exercise, your group will present the results of your experiment to the class.

Procedure

1. Be sure your graph clearly shows the results of your experiment (Make it pretty, use colors, make sure the title is informative and that the X and Y axes are labeled).
2. Write a brief summary of your experiment below and your results. In it, be sure to answer the questions below in complete sentences in a paragraph or two.
   a. What was your observation?
   b. What was your hypothesis?
   c. How did you test your hypothesis?
   d. What were your results? (Refer to the actual data shown in your graph!)
   e. What did you conclude?

Summary:

One More Thought About Hypothesis Testing

On a scientist's best days, a hypothesis is either supported or rejected by the results of the experiment. In some cases, however, the data neither supports nor refutes the hypothesis, the test is ruled inconclusive, and the scientist must start over again, making appropriate changes to the hypothesis or methods. Keep in mind that the scientific method is never ending. For this reason, scientists are very reluctant to use the word "prove." We can demonstrate a hypothesis to be false—or refute a hypothesis—and we can support a hyothesis with our results, but we can never prove it true. We can only find evidence to support the hypothesis. This is because sometime in the future, new data may arise, or a new device to test the hypothesis might be designed, that alters our understanding of the phenomenon we tested.

1 | Applying What You've Learned: The Scientific Method Lab

1. Read the scenario below and identify at least three problems with the experiment described.

   ❑ A home gardener wants to test the effectiveness of a new fertilizer. She thinks the more fertilizer she uses the taller her plants will grow but she doesn't know what the optimum amount of fertilizer is. So she sets out to test different amounts of fertilizer. She plants one plant in each of three pots that are the same type. She gives one teaspoon of fertilizer to the first plant and places it by a window in her kitchen. She gives two teaspoons of fertilizer to her second plant and places it by the sliding glass door by her back deck. She gives three teaspoons of fertilizer to the third plant and places it in her sunroom. Every day she waters the plants the same amount and at the end of one month she measures how tall each plant has grown.

2. Give an example of a hypothesis that the gardener mentioned in the scenario above could have been testing. Be sure it is testable and falsifiable.

3. Using the scenario mentioned in #1, provide an example of a possible experimental "control group" and "controlled variables" for the experiment.

4. Pretend the scenario presented in #1 was actually a well-designed experiment with the results in table 1.3. What sort of graph would you create with this data—a bar graph or a line graph? Why? Make a quick sketch of the graph below table 1.3, being sure to label the axes and to include a title.

Table 1.3.

Amount of fertilizer	Average height of tomato plants
1 tsp	8.5 inches
2 tsps	12.2 inches
3 tsps	12.4 inches

5. Having completed an experiment of your own, describe one aspect of using the "scientific method" that was surprising or different than you expected.

2	**Plant Diversity and Dichotomous Keys Pre-lab**

Use a reference book, the internet, and the dichotomous key at the end of the lab packet to answer each of the following questions.

1. Differentiate between a deciduous tree and an evergreen tree. Then differentiate between a coniferous tree and a broadleaf tree/shrub. Can a coniferous tree be deciduous? Can a broadleaf tree/shrub be evergreen?

2. On what type of plant would you find a frond?

3. Label the following structures in the figure 2.1 diagram of a branch:

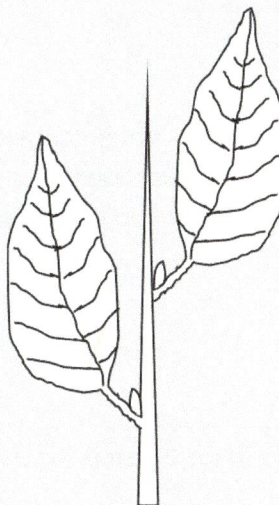

   Figure 2.1. Diagram of a Branch

   ❑ axillary (or lateral) bud
   ❑ leaf blade
   ❑ leaf margin
   ❑ leaf venation
   ❑ petiole
   ❑ stem

4. Sketch a leaf to illustrate each of the following descriptive terms:

compound leaf	simple leaf
pinnately compound leaf	palmately compound leaf
pinnate venation	palmate venation
toothed margin	lobed margin

5. While identifying plants is not a dangerous pursuit, there is the possibility of encountering poison oak. Describe the appearance of poison oak in the sun and in the shade. How can you avoid contacting it by accident?

> **You will be required to go outside for this lab. Come prepared for the weather.**

Name: _____ Lab Time: _____ Due: _____

2 Plant Diversity and Dichotomous Keys

A **key** is a tool that someone constructed to be used by others in identification of items, in our case plants. Keys may take on a number of forms. Picture-based keys rely on drawings or photographs that you look through. Picture-based keys get easier to use with practice because only experience teaches you what features in the images are essential to identification. Dichotomous keys based on descriptive terms are often easier for the novice because the author has tried to use easily observable "key characteristics." Dichotomous keys are based on these descriptions. The description comes in pairs and offers the user an either/or choice. To use a dichotomous key you have to learn what the terms mean. Including line drawings and descriptions in your key often will help you remember the terms.

A good key relies heavily on conservative characteristics of organisms, rather than labile characteristics. **Conservative characteristics** are those that show little variation between the individuals of a particular species and are unlikely to be affected by the environment. **Labile characteristics** show variation with environment or season.

For example, poison oak always has leaflets in groups of three. The leaflets always have smooth, not jagged or "toothed" margins (edges). In the sun, it grows as a short, woody shrub while in the shade it may be either low-growing or a woody vine climbing up the sides of trees. In mid to late summer it may have greenish white berries, depending on its age. In the early fall, the leaves turn yellow or red and fall off.

Q1. List two conservative and two labile characteristics for poison oak based on this description.

Basing a key only on fruits, or only on the growth form in the shade, limits its usefulness. You could not use it to identify young plants in an open field in the spring. The most conservative characteristic given in the poison oak example is leaf shape, but even then you won't be able to use the key based on leaf shape in the winter when the leaves have dropped. Often, the most conservative approach is to use a suite of characteristics that cover the seasons and environments. Most important is that you choose characteristics that can be easily seen by you and others. No key is perfect and equally easy to use at all times.

By the end of this lab, you should be able to:

- ❑ Learn the distinguishing characteristics of mosses, ferns, conifers, and flowering plants.
- ❑ Use a dichotomous key to identify living organisms.
- ❑ Explain what a species is and give examples of ways plant species differ from one another.
- ❑ Detect and identify specific detailed visible identifying characteristics of plants and be able to classify a plant as a fern, angiosperm (flowering plant), or gymnosperm (e.g. conifer) on sight.
- ❑ Distinguish between conservative and labile characteristics of species.
- ❑ Identify some common plants of Oregon's forests in preparation for the field trips.
- ❑ Learn to identify at least 10 native Oregon trees, shrubs, and ferns either by sight or by using your dichotomous key.

Exercise 2.1. Introduction to the Plant Phyla

While there are many additional phyla or divisions in the plant kingdom, we will focus our attention on four: the Anthophyta, Coniferophyta, Pteridophyta, and Bryophyta. For our purposes, we can call them, in the same order, the angiosperms or flowering plants, the gymnosperms or conifers, the ferns, and the mosses. Each of these groups has acquired an important evolutionary innovation that enables life on land and separates them from the other groups.

Early land plants, like **mosses**, developed a **cuticle** or waxy covering, to separate them from their ancestors, the green seaweeds and allow them to live on land without drying out. **Ferns** developed a **vascular system**, which enables the transport of water from roots up to the leaves and the transport of sugars from the leaves to other parts of the plant body. Each of these phyla, mosses and ferns, still have a swimming sperm and so are found primarily in wet or seasonally wet places. **Gymnosperms** or conifers escape the need for water for reproduction by developing **pollen** which is a method of delivering sperm nuclei to the ovules or eggs of plants of the same species using wind—and later, in the angiosperms using pollinators like bees. And **angiosperms** develop **flowers** and **fruits** which attract specific pollinators to transport pollen and animals to transport seeds (in fruits) away from the parent plant to minimize competition between the parent plant and its offspring.

Using the specimen before you and what you know about plants in these phyla, answer questions 2–12.

Q2. What makes a moss a moss? Examine the sample of a moss under the dissecting microscope and draw a sketch. Describe it in words. Be sure to include information about the size of the moss and its leaves. If present, note the difference between the green "leafy" part and the stick-like spore-producing part. These are technically the gametophyte and the sporophyte. Why do you think mosses are so small? What trait do they lack that other plants have? What trait do they have that allows them to live on land?

Q3. Compare and contrast a moss with a lichen. A lichen is not a plant but rather a symbiotic relationship between a fungus and an algae—and sometimes also cyanobacteria. Explain in words what you notice about the differences between their colors and structures. If present, see if you can identify several different forms of lichen (foliose, crustose, and fruticose) and moss on the same branch. Sketch what you see in figure 2.2.

Figure 2.2. Diagram of Three Basic Forms of Lichen.

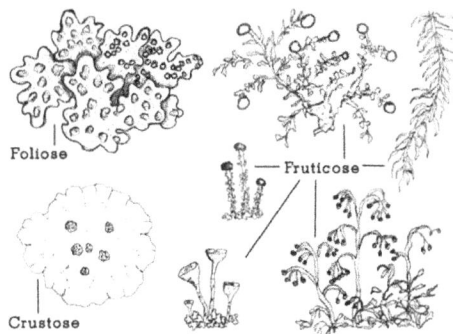

2

Q4. What makes a fern a fern? If present, examine a potted fern plant. Note that the above ground parts are called fronds. These are the fern's leaves. Where are its stems? Draw a frond then zoom in and draw one leaflet. How many leaflets are on the frond that you examined?

Q5. How do ferns reproduce? Look on the back of the leaflets. Are there any yellow-brown dots? If so, examine the leaflet under the dissecting microscope. These yellow-brown dots are called sori (singular sorus). How many sori are there per leaflet? Can you see smaller dots within each sorus? What role do you think these yellow-brown dots play in the fern's life cycle?

Q6. What sort of leaves do conifers have? Examine the samples provided for you and sketch a series of scale-like leaves, a small portion of a branch with needle-like leaves (also called awl-like), and a bundle or fascicle of needle-like leaves. Which types of gymnosperms have each of these types of leaves? Choices that we will see today are cedars, firs, pines, and spruces.

Scale-like leaves sketch	
Type of confier with this leaf type	
Needle-like leaves sketch	
Type of confier with this leaf type	
Bundled needle-like leaves sketch	
Type of confier with this leaf type	

Q7. What is the role of the cone in the life cycle of conifers? Examine the sample of cones provided in figure 2.3, pick two, and then sketch them. See if you can identify which species of gymnosperm (or conifer) the cone belongs to. Are the cones you are looking at male cones or female cones? What is the difference between male and female cones? (If male cones are not provided, see if you can figure it out by examining this life cycle below.)

Figure 2.3. Life Cyle of Gymnosperms

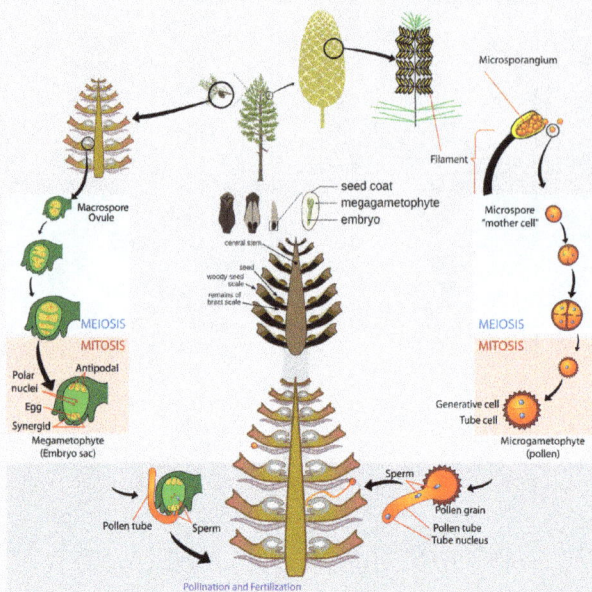

Q8. What type of leaves are found in angiosperms? You've already sketched many of these leaves in your pre-lab so think back to that and find a way to contrast angiosperm leaves with gymnosperm leaves. Describe the difference between angiosperm leaves and gymnosperm leaves in terms of their size, shape, and life span.

Q9. Why is the term "evergreen" not so useful in comparing gymnosperms and angiosperms?

Q10. If present, examine the flowers of the angiosperms. Is every flower colorful? If not, why are some colorful and others not? What is the purpose of the flower? Sketch a flower below labeling sepals, the petals, the female parts (pistil: stigma, style, ovary, and ovule) and the male parts (stamen: anther, pollen, and filament). How does fertilization happen in flowering plants? What part of the flower develops into the seed?

Q11. Is every fruit delicious to eat? If present, sketch an example of three fruits. What is the purpose of the fruit? What part of the flower becomes the fruit? Where are the seeds?

Q12. In your own words, explain how you personally will tell the difference between a moss, a fern, a conifer, and a flowering plant if you were on a walk in the woods.

Exercise 2.2. Identifying NW Oregon Plants

In this class, we will identify plants using a dichotomous key. Dichotomous means "two-branched." This type of key always gives you two choices and you choose which choice best suits the organism you are trying to identify. Remember that the descriptions may not fit exactly as any given species displays a great deal of variation between individuals depending on many factors including time of year, location of the plant, and recent weather patterns! A dichotomous key of the plants we most commonly see on our field trips is provided at the end of this lab manual. In this lab we will practice identifying common plants that illustrate the characteristics you sketched in your pre-lab using that key.

Procedure

1. Carefully remove the dichotomous key from your lab manual.
2. Follow your instructor's instructions with regards to which plants you are required to identify for this class. You may work in the lab, in the native plant garden, or in the woods.
3. Work your way through the key for each native plant specimen provided. Write down the name of the plant that you arrive at in the key.
4. Verify if you have found the correct name for your plant specimen either by checking the back of the ID card provided or asking your instructor. If you found the correct name, continue on to the next plant.
5. If you incorrectly identified the plant, go back through the key and try to figure out where you went wrong. Write down the mistake you made in the table. For example, you may have said the plant had a simple leaf when the leaf was actually compound. This is the kind of information you could write in the third column.
6. Continue practicing by using your key to identify plants around campus and in other natural areas. Remember that many of the plants used in landscaping are not native plants and may not be in your dichotomous key.
7. Remember also that you key is for trees, shrubs, and ferns. So herbaceous species like a daisy or grass will not be in your key.

If you have difficulty with this exercise, you can watch the videos available on the YouTube playlist to better understand the terminology associated with different types of plants. Scan the QR code or visit go.chemeketa.edu/plantdiversity to access the videos on this playlist including:

❑ How to make a dichotomous key Part 1
❑ How to make a dichotomous key Part 2
❑ Dichotomous key: Getting started and keying out a conifer
❑ Dichotomous key: Keying out a fern
❑ Dichotomous key: Simple vs. compound leaves
❑ Dichotomous key: Pinnate vs. palmate venation
❑ Dichotomous key: Keying out a broadleaf tree

---

8. Use table 2.1 to record the plants you have identified. Your instructor will tell you how many plants you must try to key out. Plants in the lab are labeled on the reverse side of the card and have key characteristics noted.

Table 2.1.

	My ID for this plant	Actual identity of this plant	Where I went wrong in the key and/or notable characteristic of this plant to remember next time
1			
2			
3			
4			
5			
6			
7			
8			
9			
10			
11			
12			
13			
14			
15			
16			
17			
18			
19			
20			

Exercise 2.3. Preparing for the Plant Quiz

Now that you have become familiar with a number of woody plants and ferns that are commonly found in the Willamette Valley, you can work on becoming expert in identifying some of them by sight. This exercise will be conducted largely outside of class. You can approach learning your Oregon plants a number of ways. One way would be to continue to go out in the field (woods near your house, the woods by campus, the Native Plant Garden) and use your dichotomous key to identify the woody trees, shrubs and ferns you see.

A second method you could choose along with this first approach is to make notes and drawings in the margin of your dichotomous key that give you clues to the key characteristics of each species. These notes and drawings must be done by you in your own hand. (No pasting pictures from the internet! That's not learning.)

Lastly, you can try to memorize the appearance of each species so that you can recognize each one by sight. If you choose this approach it is still important to learn the key characteristics for each species. That is, it is fine to be able to look at a tree and say, "that's a big leaf maple." And even better to say, "that's a big leaf maple because it is a deciduous tree with simple opposite leaves that have palmate venation and 5 deep lobes."

At some point during the term, your instructor will give you a quiz on these plants. The plants you will be quizzed on will most likely come from the list below. You will also be able to use your dichotomous key. Check with your instructor about the details for particular plants and the quiz format.

❑ Douglas fir	❑ Vine maple
❑ Western red cedar	❑ Oregon white oak
❑ Ponderosa pine	❑ Red alder
❑ Lodgepole pine	❑ Hazelnut
❑ Western hemlock	❑ Snowberry
❑ Sitka spruce	❑ Indian plum
❑ Grand fir	❑ Red osier dogwood
❑ Sword fern	❑ Oregon grape
❑ Deer fern	❑ Salmonberry
❑ Lady fern	❑ Thimbleberry
❑ Big leaf maple	❑ Trailing blackberry

These plants are also commonly seen in the woods around Salem. They aren't native to Oregon but are still worth knowing.

❑ Holly	❑ Himalayan blackberry
❑ Wild cherry	

Name: _____ Lab Time: _____ Due: _____

2 Applying What You've Learned: Plant Diversity

1. What is a **conservative characteristic** in a dichotomous key? Why do dichotomous keys use conservative characteristics? Re-read the intro to the lab.

2. What is a leaf? Explain how you can distinguish a simple leaf from a compound leaf.

3. Give an example of a plant **from our key** that exhibits each of the following characteristics. Do not just look these terms up on the internet.

   ❑ Tough and leathery leaves:

   ❑ Pinnately compound leaves:

   ❑ Simple palmate leaves:

❑   Leaves with sharp teeth or spines:

❑   Fronds that arise in a whorl from a central stem:

4.   In your own words, explain how you personally would tell the difference between a moss, a fern, a conifer, and a flowering plant if you were on a walk in the woods.

5.   Sketch a characteristic leaf or series of leaves of each of the following plants.

Sword fern frond	Western red cedar	Lodgepole pine
Himalayan blackberry	Oregon white oak	Oregon grape

3 | Community Interactions Pre-lab

1. Organisms interact in many ways. Complete table 3.1 by indicating whether the organism benefits (+), is harmed (-) or is not affected (0) in each of the following interactions.

Table 3.1.

Type of Interaction	Impact on Species A (+, −, or 0)	Impact on Species B (+, −, or 0)
Commensalism		
Interspecific competition		
Mutualism		
Parasitism		
Predation		

2. Which of the interactions in table 3.1 do many scientists argue is theoretical in nature?

3. What is a symbiotic relationship?

4. Which of the interactions in table 3.1 could be examples of symbiotic relationships?

5. Use the Internet to determine the basic diet of each of the following organisms and complete table 3.2. When possible provide specific examples (common names are fine) of the food these organisms consume in Oregon.

Table 3.2.

Organism	Diet
Mice	
Mole	
Pocket Gopher	
Rat	
Shrew	
Vole	

6. Differentiate between a food chain and a food web.

7. The owl pellets you will be dissecting have been sterilized in an autoclave. You will be using dissecting tools to uncover what your owl ate the day it gave up your pellet. What is one thing you can do to be safe while using dissecting tools?

3 Community Interactions

Community ecology focuses on how populations of different species interact with each other and how those relationships change over time. The interactions between organisms can be quite complex and are essential to the success of the community overall. Interactions can be classified based on their impact on the species involved in the interaction.

The most common type of interaction, **mutualism**, benefits both species involved in the interaction. **Interspecific competition** (between two different species), on the other hand, has a negative impact on both species involved. Several interactions including **predation**, **parasitism** and **herbivory** benefit one species, but harm or kill the other species involved in the relationship. Scientists also discuss **commensalism** as an idealistic interaction in which one species benefits and the other is not impacted in any way. Although many interactions have been considered as commensalism in the past, closer study of the organisms involved generally demonstrates that there is an impact on both species. It is likely that true commensalism does not exist in nature.

In this lab, we will examine several different species interactions and consider their impact on both the species involved and the environment.

By the end of this lab, you should be able to:

❑ Describe the primary modes of interactions between different species in a community.
❑ Relate these interactions to the flow of energy through an ecosystem.
❑ Hypothesize about the type of relationship that might be occurring between species based on observations of the species interacting.
❑ Build and differentiate between a food chain and a food web.
❑ Create and interpret a graph based on scientific data.

Exercise 3.1. Predator-Prey Relationships

Predation is a type of species interaction that occurs between two animals in which one animal (**predator**) kills another animal (**prey**) for food. In this relationship, the predator benefits (+) and the prey is harmed or killed (-). Predators play a key role in the balance of an ecosystem because they regulate the population sizes of their prey. Prey species tend to reproduce quickly and prolifically. When the a prey population is not regulated by a predator, their population size often increases so rapidly that they strip an ecosystem of its resources, a process that permanently alters the ecosystem and its ability to maintain a community.

The owl is an excellent subject for a study of predation because owls do not have to be trapped or dissected for us to see what they are eating. Barn owls consume their food whole and, after digesting and absorbing the nutrients, regurgitate any indigestible parts (bones, fur, feathers, etc.). The resulting **owl pellets** can be dissected and the remains within them can be identified to determine what the owl is eating.

Owl pellets can contain many things, but most of the ones we will examine will contain skulls. Any skull with teeth is a small mammal and the teeth (size, shape, arrangement) are key to their identification. The skulls you will examine most likely belonged to voles, mice, rats, or shrews. Voles, mice and rats fall into a category of mammals known as **rodents** that are classified by the presence of two pairs of permanently elongating incisors (front teeth) that are maintained by gnawing. Rodents are **herbivores** (animals that eat plants). Shrews are members of a group of **insectivorous** (insect-eating) mammals. Their skulls are characterized by the presence of sharp pointy teeth. We may also find owl pellets that contain bird skulls.

In this lab, we will dissect pellets collected from a single population of owls, determine what types of food these owls are eating and create a food web based on our class data.

Things to know

- The owl pellets we use in this lab are sterilized before they are sent to us so it is not imperative that you use gloves, but you may if you'd like to. DO NOT touch "wild" pellets without gloves.
- If you'd like to take your bones home, your instructor will give you a plastic bag. We recommend that you try to rebuild the skeleton and glue it on to a large index card for display purposes

Procedure

Materials (work in pairs)

- owl pellet
- two dissecting needles
- dichotomous key
- specimen dish
- two pairs of forceps
- squirt bottle of water
- two paper towels

1. Remove the foil from your pellet and observe it closely. Record your observations in table 3.3.

Table 3.3. Observations of Owl Pellet Before Dissection

Size of pellet (length and width; cm)	
Weight of pellet (g)	
Other observations	

Q1. What do you think you will find in this pellet? How many organisms and what type of organisms will be represented?

2. Place your owl pellet in the specimen dish.
3. Use a water bottle to moisten the pellet (do not saturate it!).
4. Use forceps and dissecting needles to gently loosen the mass of the pellet.
   a. As you find bones, place them on the paper towel.
   b. Add small squirts of water as necessary to moisten the pellet.
5. Once you have isolated all of the bones, place as much of the remaining debris as possible in the garbage, rinse your specimen dish and tools and return them to the proper location.
6. Use the dichotomous key provided to identify the number and type of species in your pellet. It can be difficult to differentiate between species of small mammals. For this exercise, we will group our organisms into major groups (voles, mice, shrews, etc.) rather than identifying them to the species level. Record your data in table 3.4.

Table 3.4. Organisms Found in Owl Pellet

Type of organism	Number in pellet

Q2. An owl generally produces two pellets per day. Based on the owl pellet you dissected, how much prey does the owl that produced this pellet consume per day?

Q3. Choose one of the prey identified in your pellet to create a food chain (not a food web) for the owl that produced the pellet in the space below.
- ❑ Refer to your pre-lab or information discussed with your instructor to identify the diets of prey species
- ❑ Arrows should be oriented in the direction of energy flow (start at the organism being consumed and pointing toward the consumer).

7. Use data from the entire class to complete table 3.5 and then calculate the percent of diet by number of food items.

Table 3.5. Contribution of Prey Species to the Diet of the Owl

Prey Type	Number found by class	% of diet by # of food items	Average weight for one individual
Mice			15 g
Mole			55 g
Pocket Gopher			150 g
Rat			150 g
Shrew			4 g
Vole			40 g
Other (bird, bat, insect, crayfish, reptiles)			—

Q4. What type of prey is most commonly consumed by the owls in your population?

8. Examine the "Average weight" column of table 3.5. Note that different prey types have different weights.

Q5. Does the weight of a prey species affect the number of prey an owl must consume to survive? Explain your answer.

Q6. If possible, talk to a group that found the skull of a pocket gopher or rat in their owl pellet. How much prey did that owl pellet contain as compared to the pellets of owls that consumed voles or mice? Does this data concur with your answer to the previous question?If no group found a pocket gopher or rat then consult the owl pellet display and hypothesize what would happen if an owl ate one of these prey items.

Q7. Use the data in table 3.5 and your research on the diet of these animals (see your pre-lab) to build a food web for the ecosystem inhabited by the owls that produced our class pellets.

❑ Fill the space provided.

❑ Use text and/or sketches to indicate the names of the organisms involved in the food web.

❑ The arrows in a food web represent the direction of energy flow. They start at the organism being consumed and point at the consumer.

❑ Remember that all food webs are based on producers. Owls, however, do not eat plants. What are the producers in your food web?

Predators can be considered generalists or specialists. A **generalist** is a predator that consumes whatever type of prey is currently available in its habitat while a **specialist** only consumes one type of prey. Consider the two graphs in figure 3.1.

Figure 3.1. Populations of Two Different Owl Species as Compared to Populations of Deer Mice Over Time

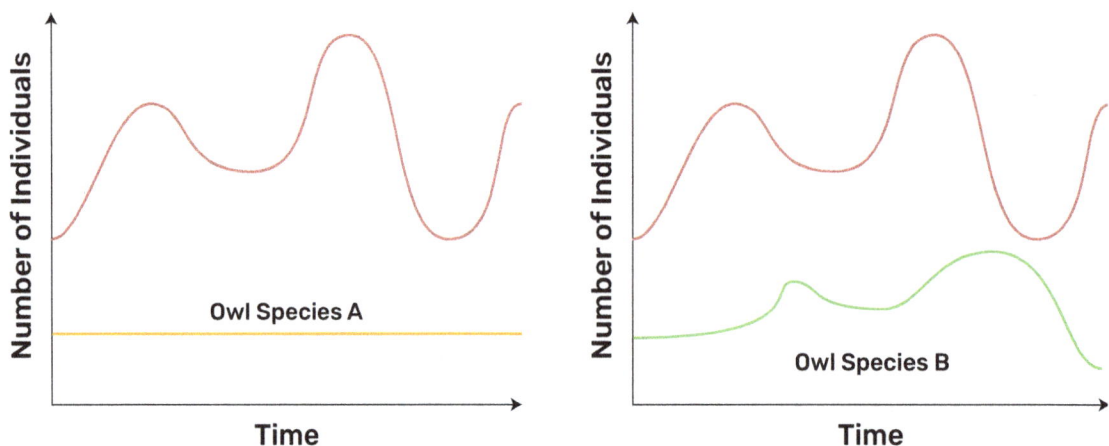

Q8. Why is the relative size of the owl population so much lower than the size of the deer mice population in both of these graphs?

Q9. Which owl species (A and B) is a generalist and which is a specialist? Use the graphs to explain your answer.

Exercise 3.2. Exploring Species Interactions

A. Symbiosis

The term **symbiosis** is used to describe an ecological relationship between organisms of two different species that live in very close association with one another and at least one of them benefits. Different types of symbiotic relationships are described based on which member benefits or is harmed by the relationship. In **mutualism**, both members gain some benefit from the relationship. In **parasitism** only one member (the parasite) benefits at the host's expense.

Here are some key pieces of information that will help you assess the symbiotic relationships we will study:

- ❏ Algae and photosynthetic bacteria, when involved in a symbiotic relationship, cannot move on their own.
- ❏ Green plants, algae and photosynthetic bacteria use light to perform photosynthesis, and make sugar. They *always* use a bright pigment (green, yellow, orange or red) to capture the light.
- ❏ Fungi, animals, protozoa and most bacteria have to get their energy by taking in energy-rich molecules that other organisms produce or store them.
- ❏ Nitrogen-fixing bacteria convert Nitrogen gas that is unusable by plants and most algae into nitrogen-containing compounds that plants and algae can use as a nitrogen source.
- ❏ Algae and photosynthetic bacteria involved in a symbiotic relationship usually have a leaky cell membrane and leak excess sugars and nitrogen compounds into their environment.

Procedure

1. Your instructor will provide 3–5 examples from the list of symbiotic relationships shown in table 3.6 for your observation.
2. Follow the procedure listed in table 3.6 for each specimen.
3. On the following pages of the lab:
   a. Sketch each specimen in the space labeled "Drawing of the Two Species Involved" on the following pages.
   b. Complete the "Analysis of the Relationship" table by listing aspects of the relationship that you think benefit or harm the participants.
   c. Write a conclusion by stating what type of symbiotic relationship is represented by the specimen. Explain how you have come to this conclusion.

Table 3.6. Specimens of Symbiotic Relationships

Specimen	Description	Procedure
Azolla (water fern)	Larger, light green cells are those of Azolla and the darker blue-green chains of cells are Anabaena, nitrogen-fixing blue-green algae.	1. Grind up a piece of specimen with a mortar and pestle. 2. Create a wet mount. 3. View with the 10X and 40X objectives.
Hydra	Hydra are animals related to sea anemones and jellys. Animals do not make chlorophyll! The green color of the Hydra comes from the green algae living inside.	Use the dissecting microscope to view the culture.
Lichen	Formed by the interaction of a fungus and an alga. The hair-like brown strands are fungal hyphae. The round green cells are algae.	1. Grind up a piece of specimen with a mortar and pestle. 2. Create a wet mount. 3. View with the 10X and 40X objectives.
Mealworm	Cigar-shaped organisms called Apicomyplexans live within the gut of the mealworm. Approximately 1/3 of all mealworms have Apicomyplexans.	1. Use a razor blade to cut both ends off of a mealworm. 2. Slice the length of the mealworm body and remove the long, dark colored digestive tract. 3. Wipe off as much adhering fat as possible. 4. Put the digestive tract on a clean slide and chop it up with a razor blade. 5. Add a drop of insect saline solution and a coverslip. Gently flatten the slide. 6. View with the 10X and 40X objectives.
Mistletoe	Mistletoe is a plant that grows on another plant. In Oregon, we find mistletoe on Gary oak and several types of pine trees.	Observe the preserved specimen provided.

Specimen	Description	Procedure
Mossy Rose Galls	Wasps lay their eggs inside the lateral bud of a rose. When the egg hatches, the feeding of the larvae stimulates the plant to grow the gall as a nutritive tissue.	1. Cut open the central core of the gall with a razor blade 2. Observe with a dissecting microscope.
Oak Galls	Wasps lay their eggs inside the leaf. When the egg hatches, the feeding of the larvae stimulates the plant to grow the gall as a nutritive tissue.	1. Cut open the central core of the gall with a razor blade 2. Observe with a dissecting microscope.
Paramecium bursaria	The clear organism swimming rapidly around the slide is the Paramecium. The green dots are algae living inside the Paramecium cell	1. Make a wet mount of the culture. 2. View with the 10X and 40X objectives. 3. Methylcellulose may be added to slow the organisms for better viewing.
Pine needle scale	White tear drop-shaped insect attached to pine needles. The insect inserts its beak-like mouthparts into the needles and sucks out the carbohydrate-rich plant juices. The scale of the female is tear-drop-shaped and showy white in color with a light yellow tip.	Use a dissecting needle to gently lift the scale while viewing under a dissecting microscope.
Termites	No animal makes the enzymes needed to digest wood, but termites live on a diet of wood. The organisms in the gut of the termite make cellulases, wood digesting enzymes. You may see conical spinning cells with long flagella (Trichonympha) and bacteria	1. Place a drop of insect saline on a slide. 2. Grasp a termite by the abdomen and gently squeeze to force a drop from its anus. Dip the drop into the insect saline. 3. Return the termite to the recovery beaker. 4. Add a coverslip. 5. View with the 10X and 40X objectives.

Specimen 1: _____

Drawing of the Two Species Involved	Analysis of the Relationship		
		Species A	Species B
	Benefit		
	Harm		
	Conclusions About This Relationship		

Specimen 2: _____

Drawing of the Two Species Involved	Analysis of the Relationship		
		Species A	Species B
	Benefit		
	Harm		
	Conclusions About This Relationship		

Specimen 3: _____

Drawing of the Two Species Involved	Analysis of the Relationship		
		Species A	Species B
	Benefit		
	Harm		
	Conclusions About This Relationship		

Specimen 4: _____

Drawing of the Two Species Involved	Analysis of the Relationship		
		Species A	Species B
	Benefit		
	Harm		
	Conclusions About This Relationship		

Specimen 5: _____

Drawing of the Two Species Involved	Analysis of the Relationship		
		Species A	Species B
	Benefit		
	Harm		
	Conclusions About This Relationship		

Q10. Are bacterial and viral diseases a form of symbiosis? If so, which type and why? Explain your answer.

Q11. Why do you think, as a rule, successful parasites do not seriously affect the health of their hosts? Think in terms of the ultimate goal of a successful parasite.

Exercise 3.3. Interspecific Competition

Competition occurs when two organisms are both using a resource that is in short supply. Competition can occur between organisms of the same species (known as intraspecific competition), but in community ecology, we focus on **interspecific competition**, or competition between organisms of different species.

In the 1930s, G.F. Gause, a Russian biologist, published a series of papers that presents what is now known as the **principle of competitive exclusion**. Gause's principle states that two species competing for the same resources under exactly the same environmental conditions cannot coexist indefinitely. Gause's classic experiment measured the population size of two different species of *Paramecium* (*P. aurelia* and *P. caudatum*) in pure and mixed cultures. These two species can easily be distinguished from each other with a microscope. Gause maintained a constant environment in the lab and changed the water and food daily. In this exercise, we will use artificial data to graph the growth of *P. aurelia* and *P. caudatum* cultures (mixed and pure) to explore the competitive exclusion principle.

Procedure

1. Obtain 4 different colored pencils.
2. Select one color to represent each pure culture (*P. aurelia* or *P. caudatum*) and one color to represent each species in mixed culture. Complete the key in figure 3.2 by indicating which color represents which culture.
3. Use the data in table 3.8 to create a line graph in the space labeled figure 3.2 showing the growth of pure *Paramecium* cultures over 30 days.

Table 3.8. Growth of *P. aurelia* and *P. caudatum* in Pure Cultures*

Day	Number of *P. aurelia* per drop of culture medium	Number of *P. caudatum* per drop of culture medium
0	6	6
2	8	7
4	10	9
6	23	20
8	35	33
10	45	40
12	55	51
14	68	65
16	79	76
18	82	78
20	86	80
22	90	82

Day	Number of P. aurelia per drop of culture medium	Number of P. caudatum per drop of culture medium
24	94	78
26	91	73
28	95	75
30	93	77

* During the experiment, the environmental conditions (temperature, light, food availability, pH, etc.) were kept constant.

Figure 3.2. Growth of *P. aurelia* and *P. caudatum* in Pure Cultures Over 30 Days

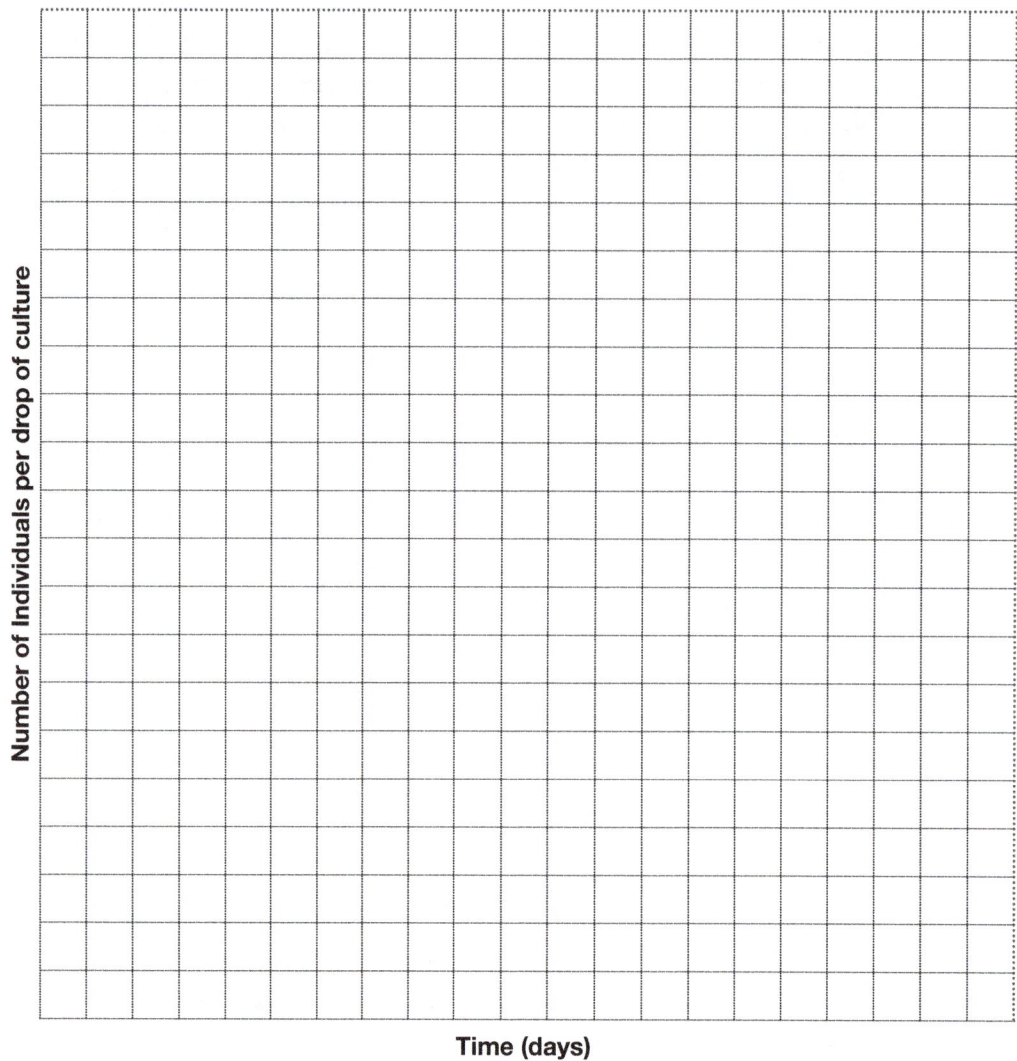

Number of Individuals per drop of culture

Time (days)

Key	
	P. aurelia (pure)
	P. caudatum (pure)

Q12. Based on your plot of the data in figure 3.2, which species is better adapted to the experimental environment? Explain your answer.

Q13. Considering your answer to the previous question, which species is likely to be most significantly impacted by competition in a mixed culture? Explain why you think this.

4. Use the data in table 3.9 to plot the growth of each *Paramecium* species in a mixed culture in the graph provided for figure 3.3.

Table 3.9. Growth of *P. aurelia* and *P. caudatum* in Mixed Cultures

Day	Number of *P. aurelia* per drop of culture medium	Number of *P. caudatum* per drop of culture medium
0	6	6
2	8	7
4	9	8
6	15	18
8	25	27
10	33	33
12	47	30
14	55	31
16	63	30
18	71	28
20	73	21
22	77	18
24	75	13
26	73	7
28	76	5
30	77	3

Figure 3.3. Growth of *P. aurelia* and *P. caudatum* in Mixed Cultures Over 30 Days

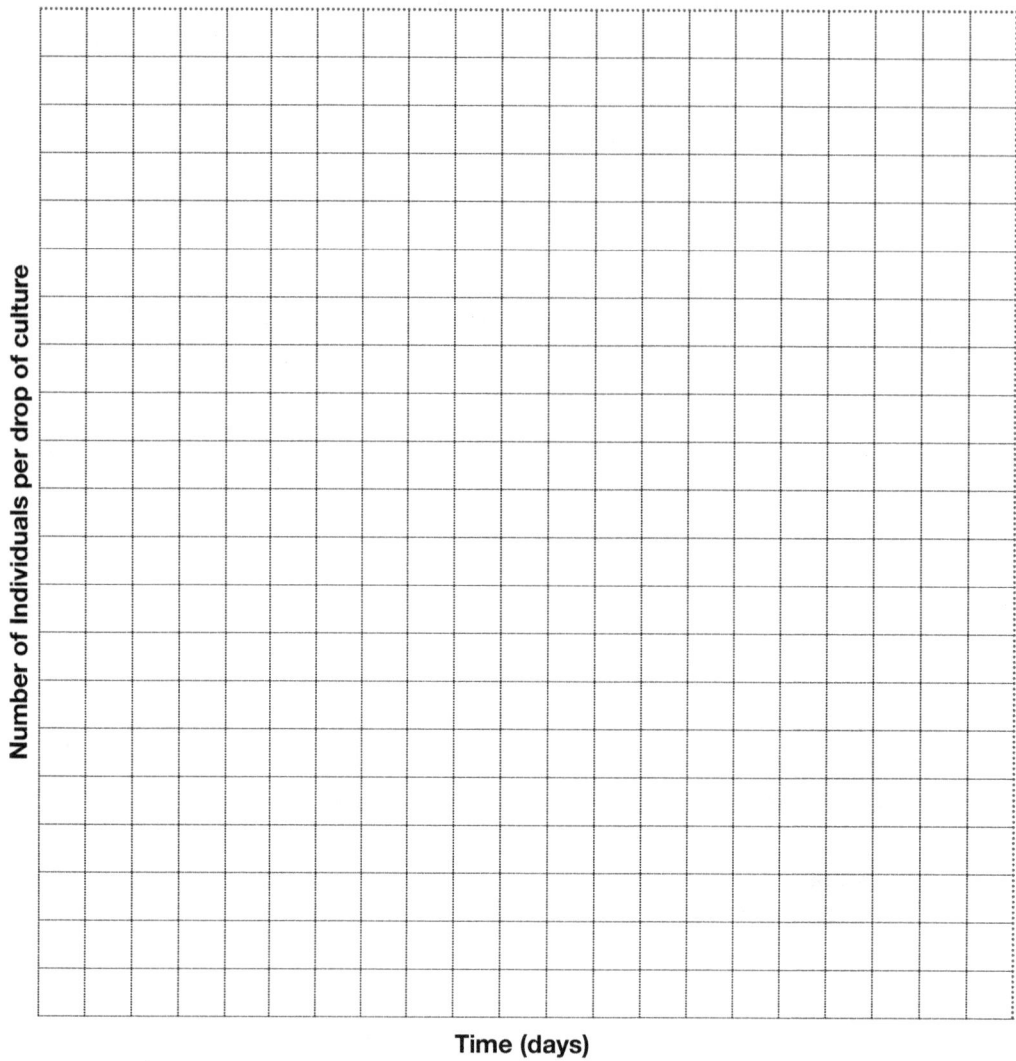

Number of Individuals per drop of culture

Time (days)

Key	
	P. aurelia (mixed)
	P. caudatum (mixed)

Q14. Is the growth curve of *P. aurelia* higher, lower, or the same when the species is grown in a mixed culture of *Paramecium* (as compared to a pure culture; compare the growth curve to the one in figure 3.2)?

Q15. What about *P. caudatum* (again, compare to the pure culture growth curve in figure 3.2)? Explain your answer.

Q16. Which species of *Paramecium* had the competitive advantage under these experimental conditions? Explain how you know.

The result of interspecific competition may be the extinction of the less successful species, but may also be a shift in the **ecological niche** of the organism. An organism's niche refers to the multi-faceted role of that species in its ecosystem. An organism's niche includes both where it lives (habitat) and how it behaves (what it eats, when it eats, when it mates, where it sleeps, etc.). Interspecific competition often leads to the natural selection of traits that reduce competition. Different species that once competed for the same resource may evolve to "divvy up" that resource, a phenomenon known as **resource partitioning**.

Q17. Are the two species of *Paramecium* used in this experiment likely to have the same niche in nature? Explain your answer.

Q18. The *Paramecium* in Gause's experiment were fed a diet of wild bacteria. An industrious general biology student repeated Gause's experiment, but fed the organisms a mixture of bacteria contaminated with Amoeba. The resulting growth curves for the organisms in the mixed culture were the same as the growth curves in the pure cultures. Can you propose a scenario that might explain these results?

Optional Exercise: Witnessing Examples of Species Interactions in Nature[*]

Using the information explained above and information you may have learned in lecture or from your book, go on a treasure hunt around campus or your neighborhood and search for examples of the types of interactions described above. Fill in table 3.7 with your examples.

> It may be easy to spot interactions of one type and more difficult to find another type of interaction. Do your best to find at least one example of each type, but as you continue your search for the other interactions, feel free to write down additional examples for interactions you have already spotted.

If you are unable to go outside for a walk, then feel free to watch a nature video to find examples of the species interactions in table 3.7. An example video is "Coral Reefs 101" from National Geographic which you can access by scanning the QR code or visiting go.chemeketa.edu/coralreefs.

Table 3.7.

Type of interaction	First example	Second example	Third example
Interference competition			
Exploitative competition			
Facultative mutualism			
Symbiotic mutualism			
Symbiotic parasitism			
Predation			
Commensalism			

[*] This exercise is completed at the instructor's discretion.

3 Applying What You've Learned: Community Interactions

1. Why are there fewer owls in a community than mice in that same community?

2. Revisit the food web you created for this lab. Food webs primarily tell us about who eats whom and consequently which direction energy flows through this web. Look more closely though and see if you can find instances of (a) competition, (b) mutualism, and (c) commensalism (thus describing an interaction web). Give an example of each from your food web in the space below.

3. Imagine a community in which you have a specialist predator and a generalist predator (owls, for example). Which species will be more affected by an illness that wipes out one particular species of prey, the specialist species or the generalist species? Why do you say this?

4. Are all symbiotic relationships beneficial to both members in the relationship? Why or why not? In your answer provide an example that supports your reasoning.

5. Explain the difference between competitive exclusion and resource partitioning in an ecological community.

Field Trip Information

What to Bring

Our field trip occurs over a lab period and takes place at a local county park. Classes with labs held on the Salem Campus will go to Bonesteele County Park. Classes from the Yamhill Campus will go to Miller Woods.

A. Prepare for This Trip

1. Complete the pre-lab as required by your instructor
2. Dress appropriately for the field trip:
    - » Long pants
    - » Comfortable shoes that can be worn off trail and in the mud
    - » Rain jacket or umbrella.
    - » Hat and gloves if the weather calls for cold

3. Bring the following items:
    - » Completed County Park Field Trip: Exploring Ecological Populations and Communities
    - » County Park Field Trip Lab packet
    - » Dichotomous Key
    - » Paper and pencil for taking notes (pens don't work in the rain!)
    - » Clipboard if you have one (some are provided)
    - » Bottle of water
    - » Snacks if needed

Field Trip Expectations

This course includes one mandatory in person field trip. As soon as your instructor informs you of the day and time of your field trip, make any necessary arrangements so that you can be present for the entire trip. To ensure that field trips run smoothly and you get the most out of your experience, we ask that you remember that you are a representative of Chemeketa Community College while on your trip. Chemeketa students and employees attending college sponsored events are expected to follow the guidelines set forth in the Student Rights and Responsibilities section of Chemeketa's Student Handbook.

Please keep the following policies in mind:

- ❑ During the field trip, students are expected to stay with the instructor and class.
- ❑ Guests (friends, significant others, children) are not allowed to attend field trips.
- ❑ Smoking is not allowed during field trips except when time and location are approved by your instructor.
- ❑ Smoking is not permitted within 50 feet of transportation or within that distance of other students.
- ❑ Consumption and/or purchase of alcohol or illegal drugs during a field trip are prohibited.

Violations of college policies will result in a score of zero for field trip participation and all other assignments associated with the field trip. In addition, your instructor is required to report such violations to the administration as described in the Student Rights and Responsibilities section of the Student Handbook.

Transportation

Because the location of the field trip is within the service area of the college, students are expected to find their own transportation to the field trip location. Talk with your instructor ASAP if this is an issue for you.

Expect the Unexpected

Although our trip to a local county park is not far from home, there still may be unexpected events occurring. For instance, even on the shorter county park trip, we may encounter bees or wasps, or poison oak. So please:

- ❑ Be prepared for delays and emergencies. While this is unlikely for our short trip to the county park, it still might make sense, if you are taking prescription medications, to bring enough for the whole day.
- ❑ Do you react to nature (bee stings, poison oak, etc)? If you have an epi pen, bring it along. Those with less severe reactions might consider bringing along an antihistamine just in case.
- ❑ Consider discussing your medical needs with your instructor and/or trip leader to ensure that we can get you the help you need as quickly as possible if an emergency arises.

Individuals with conditions that affect their balance, **have difficulty climbing or walking on uneven surfaces**, or have problems with physical exertion (e.g. mid to late term pregnancy, hip or knee surgery, serious heart conditions) should contact their instructor ASAP.

Name: _____ Lab Time: _____ Due: _____

<table>
<tr><td>**4**</td><td>**County Park Field Trip: Exploring Ecological Populations and Communities Pre-lab**</td></tr>
</table>

1. Define the following terms:

   ❑ Population

   ❑ Community

   ❑ Succession

   ❑ Seral stage

   ❑ Mature community

2. When you enter a forest, you usually find more than just trees. Instead forest habitats have several strata (or layers) of vegetation. Look through the lab and write down the lawers you should expect to see at your county park.

   ❑ What Level IV Ecoregion is your county park in? Provide the **name** of the ecoregion as well as the **number/letter designation**.

   ❑ List the names of the plants you would expect to find in this ecoregion (*Hint: there are written descriptions of each ecoregion on the back of the map.*)

3. Answer these questions after reading the Description, Characteristics, and Conservation Issues about the Willamette Valley at https://www.oregonconservationstrategy.org/ecoregion/willamette-valley/
   - ❏ Describe the size of the Willamette Valley and its climate.

   - ❏ What factors have the biggest impact on the ecological communities in the Willamette Valley?

   - ❏ Which habitat types are most affected?

   - ❏ Scan down to the photos of the strategy species. Which species are you most interested in learning more about? Why?

4. One of the questions you will try to answer while at your county park is: Is this patch of habitat a **mature** forest. Look through the lab and write down three characteristics of a mature forest.

5. When going on a field trip, one should always pay attention to your surroundings. While the dangers at our county park are limited, we may well encounter poison oak. Describe poison oak and draw a sketch below. If you brush against poison oak, what should you do when you get home?

4 County Park Field Trip: Exploring Ecological Populations and Communities

A **population** is a collection of individuals of the same species living in the same location and interacting with one another. A **species** is a group of individuals that are able to mate with one another and produce viable offspring. Populations of different species that live in the same location and interact with each other are called **communities**. And so whether you are look-ing in your own yard or a neighborhood park or on campus or in the woods, it can be difficult to separate out all of the populations you are seeing because they are all mixed together in a community. Still, there are interesting reasons why ecologists sometimes only want to study the individuals of the same species, rather than the whole community.

Examples of population-level studies include:

- ❑ Is the number of bald eagles in this region increasing or decreasing?
- ❑ How many female salmon are migrating upstream each season relative to how many males are returning?
- ❑ Does the pollen from a patch of Oregon white oak trees isolated on an island reach other Oregon white oaks on a different island?
- ❑ What is the extent of the spread of scotch broom (an invasive species)? How far west has it gotten?
- ❑ How many Canadian lynxes are left? Should it be listed as endangered?
- ❑ How many individual vine maple trees are there in each age or size class in this population?

Today we are going to begin with the study of plant populations and finish with a study of this community of plants. One question that often is asked about woody plant populations is: *How big is this population?* This seems like a simple question on its surface because we could just go out and count how many individuals there are in a forest, right? But of course, in the field, answering this question might be very difficult, particularly if the forest is very large and het-erogeneous. And so one way to begin to answer this question is to take a **subsample** of the study area by using a series of **plots**. If the area being sampled is **homogenous**, or the same throughout the area, then an ecologist might use a **systematic** or grid-like approach to lo-cating these plots. The same is true if the area has a predictable **gradient**, like an area that borders a stream on one side. For areas that are **heterogeneous**, or very different from one patch to the next, a **random sampling** design would be needed.

Another question population ecologists ask is about the **density** of a population: *how many individuals of this species are there in a particularly sized area?* We will investigate this question by using plots of the same size and tallying the number of individuals in each size class found within them.

This leads us to a third question often asked about populations: **Can this population of plants reproduce itself?** That is, are there individuals of reproductive age in the population? A related question asks: **Is the population growing over time, staying the same, or is it declining?**

One way to understand this question is to return to the location year after year and survey the same area to see if there are more or fewer of the focal species. Because we won't be able to do this for our lab, we will instead count the number of plants in different age categories that we see in each of our plots. That too is easier said than done. It can be difficult to assess a plants' age just by looking at it because some plants can hang out as seedlings for a long period of time, waiting for the right conditions to come along, that would allow the plant to grow quickly to reproductive size. And so ecologists often use size as a substitute for age.

For this week's lab we will be going on a field trip to a local forested park. There we will assess the plants that are present by examining several characteristics of plant populations, plant communities, and the process of community succession.

By the end of this lab, you should be able to:

- ❑ Identify 10 or more common Oregon plants by sight or by using your dichotomous key.
- ❑ Describe the population dynamics for one of the species present in terms of its abundance, density, and reproductive success in terms of its representation in different age/size classes.
- ❑ Conduct a community analysis of a plot to understand how your plant population is interacting with other plant species.
- ❑ Discern the state of the forest in terms of its successional pathway–that is determine if the forest is in an early successional (or seral) stage, mid-way along its pathway, or a mature forest.

Field Trip Warm-up

Your instructor will likely ask you a number of questions and engage the class in a brief discussion about the forest dynamics present in this location. You may also be asked to key out a number of plant species present as a reminder about how to use your dichotomous key and to get a sense of some of the more common species present in the location.

Exercise 4.1. Investigating Plant Population Dynamics

While you are in the forest, your instructor will provide a short lecture on **forest stratification**. When you look at many forests carefully from ground to sky, you will see that it has layers. For the purpose of this exercise, we will name the layers as follows. Complete table 4.1 to define each layer.

Table 4.1. Description in Words and Height of Each Layer of the Forest System

Layer	Description (and height in feet and meters)
Ground cover	
Herbaceous layer	
Shrub layer	
Understory layer (applies to tree species)	
Canopy layer (applies to tree species)	

Using the information above, you are now ready to study one of the many plant populations around you. Your instructor will assign you a particular plant species to investigate that you identified in Exercise 1. Be sure you know what this species looks like before you head off to sample your plots.

Procedure

1. Obtain a method of measuring your plots from your instructor. This will likely be a string and some flags but could be a measuring tape or a quadrat.
2. Also find out from your instructor which plant species you are investigating.
3. Locate your first random plot by having one person in your team of two or three people pick a random number between 25 and 150 and another person pick forward or backward. Then have a different person pick a second number between 5 and 30 and another person pick left or right. These are now the number of steps you will take from your current location.

PLEASE NOTE: Your instructor may have a different method for you to find your plots. The most important thing is that you use a method that eliminates your bias in picking a plot. That is, you shouldn't pick a plot because it has a lot of your species in it, or because it doesn't have any blackberries in it. Use the first number for the number of steps you take along the path; use the direction (left or right) to determine which direction you will move into the forest; use the second number as the number of steps you will take from the path into the forest to locate the center of your plot.

4. When you arrive at the location of your plot, use the flags and string provided to mark the boundary of your plot. Plant one flag in the center of the plot. Have one person stand with one end of the string over the flag. Have a second person walk away from the first person holding the other end of the string. Then, as best as you are able, have the second person walk in a circle around the first person, marking the circumference of the circle with additional flags (if provided) or with mental notes about what is "inside" and what is "outside" the plot.

5. Once in your plot, identify the number of individuals of your species that are present in each size class (ground cover, herbaceous layer, shrub layer—and if you have a tree species, understory and canopy.) Fill this information into table 4.2.

6. Repeat steps 3–5 two more times, locating new plots to study the same species that you were originally assigned.

7. Gather with your classmates and instructor to discuss the data collected.

Table 4.2. Occurrence of a Particular Plant Species in Each Layer of the Forest System.
Write the name of your species in the top left corner.

Plant Species	Plot 1 Number of individuals	Plot 2 Number of individuals	Plot 3 Number of individuals	Average Number of individuals
Ground layer				
Herbaceous layer				
Shrub layer				
Understory layer (for trees)				
Canopy layer (for trees)				
Total in all size classes				

In the area we are sampling, there are approximately 150 plots of the size we were using today.

Q1. Using this number of plots, estimate the total number of individuals of your species in this forest.

Density is a measure of the number of individuals in a given area. Sometimes, the area unit used is one square meter for small plants, and sometimes, the area might be a square kilometer for larger plants. When you calculated the average number of total individuals in all size classes above, you were basically also calculating the density of your species "per plot." However, our plots are not a standard unit of measure.

Q2. The size of the circular plot we have been using is 78.5 m² (Area of a circle = πr^2). What is the density of your plant species per 1m²? (Your answer will likely be a decimal, and it might be a very small decimal.)

Q3. Is your population of plants able to reproduce itself? What evidence do you have from the data you collected to support your claim?

Q4. Based on your answer above, is the population size in the future likely going to increase, decrease, or stay the same? Explain your reasoning.

Exercise 4.2. Understanding Community Succession

A. Which Stage Is This Forest In?

Ecological succession is the way in which an ecological community changes over time. As you can see in the picture below when a natural community is disturbed in some way, whether by something human-caused, like farming, or something natural, like a hurricane, the effect is like resetting the community back to an earlier stage of development. Each stage of the process is called a **seral stage**. The earlier stages are referred to as **pioneer communities**, and the later stages are called **mid-successional**, **late-successional**, or **mature communities**.

You may have learned the term "climax community" in a previous class. This term is similar to the idea of a mature community and is not used by ecologists so much anymore because it connotes an endpoint or highpoint of the community. More recent understanding of community dynamics recognizes that plant communities are often more cyclical in nature, with small disturbances frequently occurring within a community and large disturbances occurring less frequently but still as natural parts of the cycle. When you think about communities this way, it becomes clear that there is no endpoint, or "climax" for the community.

Figure 4.1. Sketch of the Stages of Forest Succession. 1. Bare rock. 2. Mosses and grasses. 3. Grasses and forbs. 4. Shurbs added. 5. Mixed conifer and hardwood species. 6. Hardwood species dominate canopy.

Determining the ecological successional stage of a forest can be difficult. Each type of community will have different species present in its mature stages, so comparing a mature forest in North Carolina with a mature forest in Colorado with a mature forest in Oregon is very much like comparing apples, oranges, and bananas. And yet, there are some characteristics of a "mature forest" that ecologists can apply to any forest or plant community.

For a forest community, these characteristics include:

1. Well-developed soil
2. A high diversity of species (for the region)
3. Representatives of the tree species that are found in the canopy layer also in the other age classes (as seedlings and saplings)

This last criteria, (#3), means that whatever species are present in the forest canopy are able to reproduce themselves in future generations. If the trees in the canopy do not have seedlings surviving in their own shade, this generally means that this species is a sun-loving species, and so will eventually be outcompeted by tree species that *can* grow seedlings in their own shade. Another way to think of this is if a tree species is not breeding now—if it doesn't have seedlings now—it probably will not be present at all 50 or 100 years from now.

If you refer to the diagram in figure 4.1, you will see that it would take a long time indeed to actually follow the successional pathway of a forest in real-time. From abandoned fields to mature forests would take longer than several human lifespans. And so, as a substitute for this, we can gather all the population data from each team and study the patterns you find there. You can use table 4.3 to see which species have representatives in each size class layer of your community. Your instructor may provide additional information from this study* of the plausibility of forest restoration at Bonesteele County Park. One key take-home message from this publication is that the typical successional pathway at Bonesteele is:

**oak savanna → oak, douglas fir, bigleaf maple, grand fir →
big leaf maple and mixed hardwoods**

---

*   Scan QR code:

Q5. In terms of ecological succession, what would it mean if the canopy species present DO NOT have seedlings or saplings growing up underneath them? What would it mean if they DO?

Procedure

Your instructor will likely facilitate a conversation with the whole class about which species different groups found in each layer of the forest. Alternatively, you may need to circulate among your peers to get their data.

1. Fill in table 4.3 with the names of the species present in each layer. For instance, if Grand Fir were found in the canopy, understory, and shrub layers, you would write "Grand Fir" in each of those boxes but not in the herbaceous layer or ground cover layer. Remember from the introduction that these forest layers are being used as a substitute for age.

Table 4.3. Cumulation of Class Data for All Plant Populations Examined

Forest Layer	Species Present
Canopy	
Understory	
Shrub Layer	
Herbaceous Layer	
Ground Cover	

2. After your group discussion, answer questions 6–11, either in the field or as homework.

Q6. Did you find plants in all four layers of the forest (herbaceous layer, shrub layer, understory, and canopy)? If not, which layer was lacking?

4

Q7. Were there any individual species that were present in three or four of the layers? If so, what species are they?

Q8. Were there any individual species that were present in two layers? If so, which species are they?

Q9. Which plant species were most common? Which plant species were rare (you only saw 1–2 individuals of this/these species)?

Q10. Do you appear to be in a healthy forest or was it overrun with invasive species?

Q11. Using evidence from table 4.3 and questions 6–9, answer the following question: is the forest in this county park a mature forest or is it still in an earlier stage of succession (a seral stage)? What is your reasoning for your answer?

Exercise 4.3. "Ground Truthing" the Class Data Set and Taking the Plant ID Quiz

Now it's your turn to verify if the combined data about each plant population in your county park actually represents what is going on with the plant community here. Return to one of your plots and conduct a full community analysis. This will also allow you to practice identifying more of the plant species. During this time, your instructor will give the plant ID quiz either to small groups of students or to the whole class, depending on the time remaining.

Procedure

1. To conduct the community level analysis, list the species you identify within your circular plot in the first column of table 4.4 and put a check mark in each size class column that you find the species.

Table 4.4. First Plot for Community Analysis of Plant Species Present in Different Layers of the County Park Forest

Species ID	Herbaceous layer < 1m tall	Shrub layer 1–5 m tall (~ 15 ft)	Understory layer >5 m < canopy	Canopy layer Tallest trees

2. Now move to a second area of the forest to conduct a second community level analysis in a different area. List the species you identify within your second circular plot in the first column of table 4.5 and put a check mark in each size class column that you find the species.

Table 4.5. Second Plot for Community analysis of Plant Species Present in Different Layers of the County Park Forest

Species ID	Herbaceous layer < 1m tall	Shrub layer 1–5 m tall (~ 15 ft)	Understory layer >5 m < canopy	Canopy layer Tallest trees

Q12. Does the data you collected on all the species present in your plot agree with the combined data from the class plant populations? In what ways does your ground truthing seem to support the class data? In what ways is it different?

Table 4.6. Plant Quiz Answers

	Plant name
1.	
2.	
3.	
4.	
5.	
6.	
7.	
8.	
9.	
10.	
11.	
12.	
13.	
14.	
15.	

Name: _____ Lab Time: _____ Due: _____

<table>
<tr><td>4</td><td>Applying What You've Learned: County Park Field Trip: Exploring Ecological Populations and Communities</td></tr>
</table>

After returning from the field trip, answer any questions that you left blank while at your county park. As you go back through the questions, reflect more deeply on what you learned. You have hopefully come away from the experience with a greater understanding of the facts of plant identification, plant population size and density, plant community succession, and plant diversity. Additionally, you may have learned something about conducting field work, collaborating with other researchers, and observing details.

For this week's AWYL, **create a well-composed graph or figure based on the data** that you collected today. Think back to the Scientific Method Lab. What sort of data did you collect? What sort of graph is best for this type of data? Could you represent what your data told you more coherently in a diagram instead of a graph? Be sure to give it a descriptive title and, if it is a graph, label the axes and provide a key if necessary.

Then, **on the back side of this sheet**, write a short paragraph (7–10 sentences) that interprets the information you presented in the graph or diagram, demonstrates your new found knowledge about populations and/or communities, and reflects on your experience in the field. Finally, answer the main question of this field trip using evidence that you and your peers gathered: Is the forest in your county park a mature forest? Why or why not?

Graph or Diagram

Summary Paragraph

4

5 | The Human Population Pre-lab

This is an easy pre-lab. You will not be graded on whether or not your answers are right, just on whether you completed the work. **Do not research these answers.** Just write down your best guess. You will check your answers on your own later during lab.

1. Which of these is closest to the current human population of the world?
   - ❑ 200 million
   - ❑ 400 million
   - ❑ 1 billion
   - ❑ 8 billion
   - ❑ 10 billion

2. Rank the following world regions in terms of human population size today and the projected human population size in the year 2100. Label the region with the largest population "1," and the next largest with "2" and so on. Then make the lines into two boxes or a table with Present and 2100 at the tops of the columns and the regions as the rows.
   - ❑ North America _____
   - ❑ Latin America _____
   - ❑ Asia _____
   - ❑ Europe _____
   - ❑ Africa _____
   - ❑ Oceania (Australia, New Zealand, S. Pacific Islands) _____

3. What is the approximate current human population of the U.S?
   - ❑ 3 million
   - ❑ 43 million
   - ❑ 340 million
   - ❑ 630 million
   - ❑ 850 million

4. How is the population of the U.S. changing (ignoring the effects of immigration)?

- ❏ growing quickly
- ❏ growing slowly
- ❏ growing at a moderate rate
- ❏ not growing at all
- ❏ shrinking

5. How is the human population of the world changing?

- ❏ growing quickly
- ❏ growing slowly
- ❏ growing at a moderate rate
- ❏ not growing at all
- ❏ shrinking

6. List the three most populous countries (not continents) in the world:

- ❏ _____
- ❏ _____
- ❏ _____

7. What is the average life expectancy of a person in

- ❏ the world?_____
- ❏ the U.S? _____
- ❏ Japan?_____

8. Which countries in the world have the biggest ecological footprint per capita? Which countries have the smallest ecological footprint per capita?

Name: _____ Lab Time: _____ Due: _____

5 The Human Population

Recall from our study on populations that population size and density are among the easiest to measure at a given point in time. When we consider the human population, we find these factors are affected by fertility rates, birth and death rates, migration rates, mortality, survivorship, life expectancy and carrying capacity. The study of all these factors together is called **population demography**.

Due to technological advances, the population of humans on Earth has changed dramatically in the 200,000 years that we have existed on the planet. The human population has transformed from small bands of hunter-gatherers, to larger groups of sedentary farmers, to enormous groups of industrial and post-industrial workers. Throughout this process, humans have transformed the world to that point that the population of humans could now be too large for our biosphere. Many factors indicate that humans have exceeded our carrying capacity and our behaviors are affecting our ability to coexist with other populations on our planet.

And yet, recent research demonstrates that the size of the human population is beginning to level out. How we, as a global human population, address our own population size and our interactions with the rest of the biosphere will determine what the carrying capacity is for humans on this planet.

In this lab, we will investigate the growth of human populations regionally and globally and explore the implications of that growth on future population size and the surrounding biosphere. As you work through the lab exercises, please keep in mind that we are more interested in your understanding *concepts* than in your memorizing numbers. We will then apply the concept of carrying capacity to the human population.

By the end of this lab you should be able to:

❑ Discuss current and future trends in the world's human population.
❑ Define terminology associated with population dynamics; for example "total fertility rate," or "replacement rate," or "population momentum."
❑ Understand and discuss the implications of different age structures in a demographic pyramid on future population size and on maintaining cultural and societal norms.
❑ Relate current and future populations size to resource use and carrying capacity.
❑ Determine your own ecological footprint and discuss the what that means in terms of earth overshoot day.
❑ Examine the impact of humans on Chemeketa's campus.

Exercise 5.1. Exploring Demographics of the Human Population

To begin this lab, we will explore information about the human population available to us through the GapMinder Foundation founded by Hans Rosling. Hans Rosling (1948–2017) was a Swedish physician, demographer and professor of International Health at Karolinska Institutet in Sweden. He worked in Africa for over 20 years and made a series of TedTalks about the current state of the global human population and our misconceptions about other regions of the world. Rosling argues that as long as these misconceptions about the developing nations exist in the mindset of people in developed nations, we will continue to create policies that are mismatches for the challenges we face together globally.

In 2013, Rosling created an hour-long documentary, called *Don't Panic—The Truth about Population*, that brings together many of his ideas from his previous TedTalks. Each of Rosling's videos is available to stream from the GapMinder Foundation website. We will watch a short excerpt from one of the videos that explains human population growth projections to the year 2100. Visit go.chemeketa.edu/populationgrowth or scan the QR code to access this documentary. Start this video at minute 19:15 and end at 28:20.

Exercise 5.2. Exploring the World Population Data Sheet

For this week's pre-lab, you guessed the human population. Use the World Population Data Sheet to check your answers, your instructor may have paper copies. You can also access it via the QR code link. After reviewing your pre-lab and misconceptions, answer the remaining questions in this exercise.

Q1. Were you very far off on any of your predictions from the pre-lab? Did any of the correct answers surprise you? If so, list them:

Q2. If you were incorrect for any of the answers in the pre-lab, why do you think you were? What was the source of that information that ultimately led you to make incorrect guesses?

Q3. How can we, as individuals and as a society, get beyond our misconceptions?

A. Birth Rates

Next, we'll examine births and fertility rates around the world on the World Population Data Sheet.

Procedure

1. Look at the "Births per 1000 individuals" column. Quickly scan the column and answer the following questions.

Q4. List 3 countries that have among the highest birth rates. You don't have to tediously search for the 3 with the absolute highest birth rates—just get a quick sense. On what continent(s) are these countries located?

Q5. List 3 countries that have among the lowest birth rates. Again do this after a quick scan of the data sheet. On what continent(s) are these countries located?

Fertility rate refers to the number of children born per reproductive female. **Replacement rate** refers to the number of children a woman should have to ensure no change in population size. Remember that the woman is not just replacing herself, but her mate as well.

Q6. How are fertility rate and replacement rate related? For a population to remain the same size, what should be true about the fertility rate and the replacement rate?

Q7. You would think to replace herself and her mate in the next generation, a woman would need to have 2 children. However, the replacement rate is actually 2.1 per reproductive woman. Why is it 2.1 instead of 2.0? (This is a thought question and the answer is not on the data sheet...)

Table 5.1 contains information from the World Population Data Sheet. Use these data to answer questions 42–46.

Table 5.1. Data from the World Population Data Sheet

Continent	Fertility rate (births per female)	Infant mortality rate (per 1000 born)	Life expectancy (in years)	
			Males	Females
World	2.5	36	70	74
Africa	4.7	57	59	62
Americas	2.0	14	74	80
Asia	2.1	31	71	74
Europe	1.6	5	75	81
Oceana	2.3	20	75	80

Q8. Examine table 5.4. To see if there is a relationship between infant mortality rate and fertility rate, sketch a quick graph below with infant mortality rate on the horizontal axis and fertility rate per female on the vertical axis.

Q9. Use figure 5.1 to describe what is meant by a positive relationship between variable 1 and variable 2, a negative relationship between the two variables, and no relationship between the variables.

Figure 5.1. Relationship Between Variables.

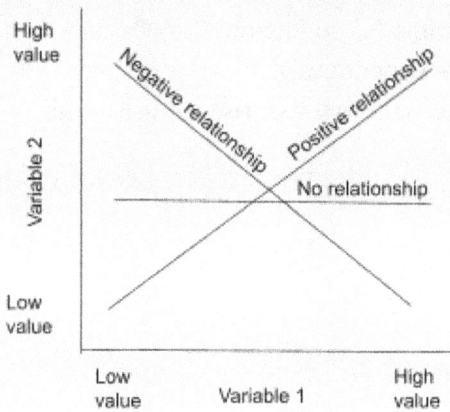

Q10. What sort of relationship do you see between infant mortality rate and fertility rate? Then explain what happens to the fertility rate as the infant mortality rate goes down.

Q11. In which continent(s) is the fertility rate greater than 2.1? What does this imply for population growth on this (these) continent(s)?

Q12. In which continent(s) is the fertility rate lower than 2.1? What does this imply for population growth on this (these) continent(s)?

B. Exploring the Data That Goes into an Average

Averages can be both informative and misleading. Work through the two examples below to better understand the value and limitation of presenting a data set only by its average.

Procedure

1. Find the average of the two populations in table 5.2 to explore the effect of high infant mortality on the average age at death (or life expectancy).

Table 5.2. Comparison of the Average Age at Death (or Life Expectancy) for Two Made-up Populations

	Population 1 = High Infant Mortality	Population 2 = Low Infant Mortality
	1	2
	2	18
	3	36
	5	36
	18	42
	36	54
	42	54
	54	66
	66	66
	72	72
Total		
Average		

Q13. What is the effect of more people dying in early childhood on life expectancy?

Q14. In populations with high infant mortality, can some people still grow old? Provide evidence for your reasoning. (*But don't overthink it.*)

Q15. Does presenting the averages for these two data sets convey enough information about these two populations? Why or why not?

2. Now, take the averages of the life expectancy data from the two made-up human populations in table 5.3 and sketch a quick graph of this data next to the table. What type of graph would work best for the data?

Table 5.3. Comparison of the Average Age at Death (or Life Expectancy) for Two Made-up Populations

	Population 1	Population 2
	1	36
	3	37
	5	38
	25	39
	50	40
	55	41
	60	42
	65	43
	70	44
	75	45
Total		
Average		

Q16. What is the average for each population? Do these averages adequately represent the age at death (life expectancy) in the two populations? Why or why not?

Q17. What could you add to your graph to have it better represent the range of ages at death for the two populations?

Q18. Summarize in a sentence or two very clear bullet points what the main advantage and disadvantage is in reporting the average of a data set.

Q19. Apply what you have learned about averages to the Hans Rosling *200 Countries over 200 Years* video. What are the important takeaways from the video, and what are the limitations?

2.   Now we will watch an explanation by Rosling from the longer video, *Don't Panic: The Facts about Population*, to understand the projection of global human population growth rate to 2100. Scan the QR code or visit go.chemeketa.edu/prbdatasheets to access this documentary. Start this video at minute 19:15 and end at 28:20.

Q20. When will population growth slow down? How many people are predicted to be in the world at that time?

Q21. Think about Rosling's story about his granddaughter. How many children were in the world in the year 2000? How many children will there be in 2100? What is meant by "peak child?"

Q22. How can the world population continue to increase to 11 billion by 2100 if the number of children will remain at 2 billion from now until 2100? Consider Rosling's foam block exercise. What is the "inevitable fill up of the adults?" (Note: this is also called population momentum.)

Q23. In today's world population, where do the billions of people live? In 2050? In 2100? What is the "pin code of the world?" Fill in table 5.4 with your answers.

Table 5.4. Billions of People in the World by Continent

Continent	Today	2050	2100
Americas			
Europe			
Africa			
Asia			

C. Age Structure in the Population

To gain a better understanding of a population within a nation, it is important to examine how that population is distributed across different age groups. In the broadest sense, one can ask, are there a lot of old people in this country? Are there a lot of young people? The **age structure** of a population (or number of people in different age groups) can indicate many things about the future size of the total population.

The age structure also has implications on many societal questions in a population, such as: Will there be enough jobs for everyone? Will there be enough social services? Who will take care of the old people? Who will run the industries/farms/government?

Procedure

1. To gain a better understanding of the age structure in different countries around the world, look over the columns for Percent of population ages <15 and Ages 65+ on your World Population Data Sheet. This data has been added to table 5.5 for a few select countries for you to analyze.

> This data will also help you analyze the population pyramids in Exercise 5.2.D.

Table 5.5. Age Structure Information from World Population Data Sheet

Country	Percent of population ages < 15	Percent of population ages 65+	What does this mean to the size of this population in the future and what societal challenges will a population with this age structure face?
India	29%	6%	
USA	19%	15%	
Kenya	42%	3%	
Japan	13%	27%	
Russia	17%	14%	

D. Understanding the Complete Population Age Structure Pyramid

In table 5.5 you compared the percentage of young people in a population to the percentage of older people in a population. In figure 5.2, you will find the 2023 population pyramids from the website Population Pyramid. Scan the QR code or visit go.chemeketa.edu/populationpyramid to access the website.

Procedure

1. These population pyramids have the 2023 data for all age groups in the world and in Russia. Examine each pyramid closely and answer the questions below.

Figure 5.2. Population Graphs

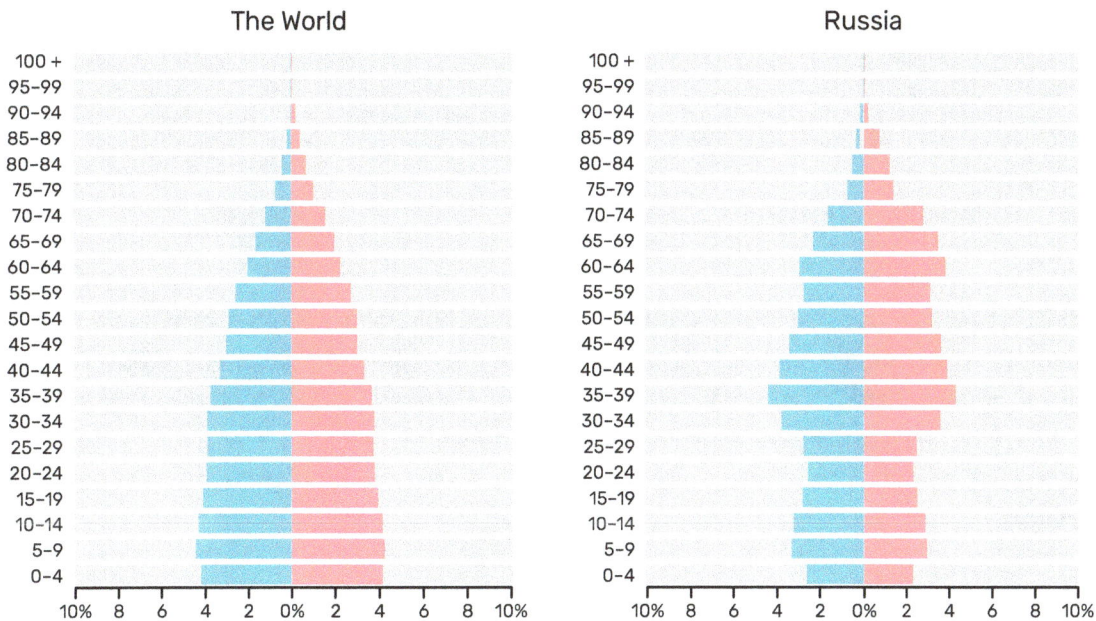

Q24. Compare Russia to the World. What percentage of the population is males in the 55–59 age group? What percentage is females in the 20–24 age group? What is the total percentage of babies and toddlers (0–4 years old, males + females) in the world and in Russia? Which is higher?

Q25. Apply what you learned about averages earlier and estimate the average life expectancy for males and females in Russia. For which group is the average likely higher, males or females? Why do you say this? (Note: average life expectancy is different from the oldest possible age.)

Q26. Considering your answers to the questions above, what can you conclude about the Russian population as compared to the World population both now and in the future?

Now that you are good at interpreting data from a population pyramid graph, see if you can match the population pyramids in figure 5.3 A–D to the <15 or >65 data in table 5.5. The pyramids in figure 5.3 are 2023 data from the Population Pyramid website for India, Japan, Kenya, and the US. The names have been removed, but remember you can refer back to the data in table 5.5. Once you match the country to its data, write the name of the country below the pyramid.

Q27. How does the shape of each of these pyramids relate to its future population size?

❑ A triangle-shaped pyramid indicates that the population will _____ in the future because:

❑ A rectangular pyramid indicates that the population will _____ in the future because:

❑ An ice cream cone-shaped pyramid indicates that the population will _____ in the future because:

Figure 5.3. Population Pyramids A–D

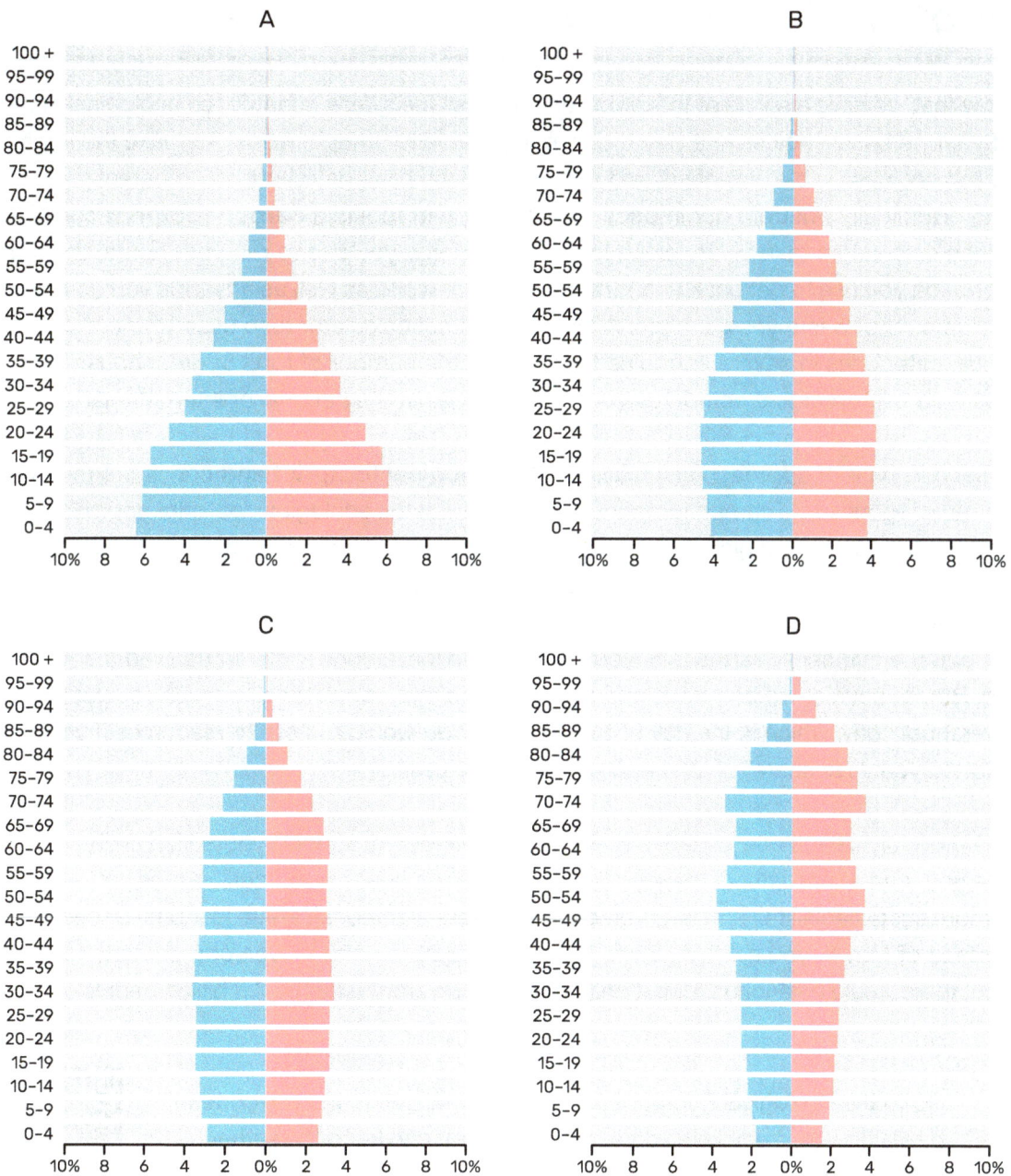

Q28. What is the shape of the population pyramid for the US? And so what are your expectations for the future population in the US without immigration?

Exercise 5.3. Understanding the Ecological Footprint of Individuals

Now that we have a sense of the global use of fossil fuels, let's quantify the ecological impact of your personal choices in a way that is comparable to other humans. To make the impacts of our choices visible, scientists and social scientists have devised a few different tools that can graphically illustrate our impact. The first type of tool is a **carbon footprint** calculator. These calculators take a very close look at the energy you use to run your home, the transportation you use to get back and forth to school or work, as well as any trips you go on, and the amount of waste you generate. A carbon footprint calculator then uses your estimates of these categories in several formulas to calculate the number of tons of greenhouse gasses you emit each year and compares your number against the national or international average.

A second type of approach calculates your **ecological footprint** and considers much the same categories as a carbon footprint calculator. With the ecological footprint approach, your impact is calculated in global hectares, or the amount of land and water that must be transformed in order to generate the fuel, food, and fiber that you use. The graphical representation of an ecological footprint in the calculator we will use in this lab is the entire earth. Or, better put, the number of earths that would be needed if everyone in the world lived the way you do. But, of course, we only have one earth. And so this calculator also gives you your "earth overshoot" day, which is the day of the year that you would run out of resources if you were allotted your "fair share" of land and water for one year.

It is probably clear that a lower number of earths needed to support yourself, if everyone lived like you, and a date later in the year before you've run out of resources, means you are living a more sustainable life. That is, if you can make your allotted fair share last until December 31 instead of May 8, you are living a life that is sustainable—and if everyone on earth lived like you, there would be enough food, fiber, and fuel to go around for the 7.8 billion people on earth.

Procedure

1. Go to the Ecological Footprint Calculator by scanning the QR code or by visiting go.chemeketa.edu/ecofootprint. You can keep the settings at the basic slide bar or click "Add details to improve accuracy" if you would like. Answer the following questions.

> You will need to enter an email address to do this exercise, but you will not receive any emails from this organization. When you finish, be sure to click the "See details" arrow and the "Explore Solutions" button to answer the questions below.

Q29. How many earths would be needed for the world population, if everyone in the world lived just like you?

Q30. What does the website mean by "earth overshoot day?"

Q31. What is your personal earth overshoot day? Are you surprised by this date? Explore the solutions section at the end of your information and describe two suggestions that would help you (and everyone) conserve resources further into the year.

Q32. Add your data to the graph started by your instructor. What is the range of earths needed if everyone on earth lived like us? Where is the peak?

Q33. The Ecological Footprint website measures the impact of people's resource use in global hectares (Gha). How big is a hectare in terms you can visualize (acres or football fields?) For which area of consumption did you need the highest number of global hectares? The lowest?

Exercise 5.4. Assessing Human Impact on Campus

Now it's your turn to measure the impact of humans on the earth. As a class we will try to determine the different ways that we have transformed campus. Working in pairs, we will be using a modified version of a point-transect system. You and your partner will be given a 5m length of rope to use as your transect. Each team will be assigned a different part of campus to evaluate. Teams may be assigned to start near the same centerpoint and move outward like the spokes on a bicycle or they may be assigned different parts of campus to assess, depending on which campus you are assessing. Your instructor will tell you where to start. Once there follow the directions below and fill out your data sheet.

Procedure

1. Find the starting point for your transect.
2. Be prepared to measure or pace off 200m in a straight line (so be sure you have enough room for this distance from your starting point.) Note: you should pick the end of the transect before you start and keep walking toward that endpoint no matter what is between your starting point and the end point (within reason, please do not ruin a flower bed or other sensitive area.)
3. Record the substrate at your starting point. And at 5m to the right and 5m to the left.
4. Then "leapfrog" your partner so that you measure out 10m with your 5m rope. Record what substrate you encounter at 10m and again 5m to the left and 5m to the right of this spot.
5. Continue to record what your foot encounters every 10m and 5m to the right and left until you reach the 200m mark. Note: you should not be looking all around you to record what you find, but literally what type of substrate your foot touches.
6. If your transect runs into a building, estimate how many of your data points would be "building" to get to the other side of the building and resume your transect in as straight of a line as you can.
7. Record all this information in the spreadsheet below.

Substrate Options
- ☐ Grass
- ☐ Dirt
- ☐ Flower bed
- ☐ Bark mulch
- ☐ Weeds
- ☐ Cement (sidewalk)
- ☐ Asphalt (road)
- ☐ Building
- ☐ Native plants
- ☐ Other

A. Hypothesis About Campus

(This could be about how extensive the human transformation is or what type of substrate you think will be most prevalent.)

Table 5.6.

Meters along transect	Substrate at this location	Substrate 5m to left	Substrate 5m to right	Other notes
0				
10				
20				
30				
40				
50				
60				
70				
80				
90				
100				
110				
120				
130				
140				
150				
160				
170				
180				
190				
200				

5

B. Notes and Observations About Your Transect

Q34. Where was your transect?

Q35. What type of area is it in?

5

Q36. How extensive are the human impacts here (high, medium, low)?

Q37. What seems to be the biggest human impact?

Q38. What is the smallest human impact?

Q39. At this stage of your data collection and analysis, do you think your hypothesis will be supported or refuted? Why?

2. When you return to your lab room, tally the number of observations you have of each substrate in table 5.7. You should have 63 observations (3 data points every 10 meters for 21 points).

Table 5.7.

Substrate	Your total	Class total	Percent of total that is this substrate
Grass			
Dirt			
Flower bed			
Bark mulch			
Weeds			
Flower bed			
Cement			
Asphalt			
Building			
Native plants			
Other			
Total			

C. Data Analysis

1. Create a graph of the class data.
2. You can choose between creating a bar graph or a pie chart. You will need to justify your choice below.
   a. To analyze the substrate data as a bar graph:
      i. Tally the number of observations observed for each substrate
      ii. Make a graph of the number of times you encountered each type of substrate.
   b. To analyze the substrate data as a pie chart:
      i. Tally the number of observations observed for each substrate.
      ii. Divide the number of observations for one substrate by the total number of observations to get the percent of this substrate of the total substrates observed. (e.g. if you stepped on grass 43 times and had 80 observations, then you would divide 43 by 80 and get 0.5375, or roughly 54% of the observations are grass.)
      iii. Repeat i-ii for each of the substrates
      iv. Make a pie chart for this area.
3. Be sure to label your graph with a descriptive title and label the axes or pie pieces as appropriate.

D. Data Interpretation

1. Now that you have your data analyzed, you need to summarize what you found.
2. In your interpretation (called the "discussion" in a scientific paper) be sure to address the following questions:
    a. What was the big question you were trying to answer?
    b. What was your hypothesis?
    c. What did the data show? (Give some of the highlights of the data using the real numbers)
    d. Was your hypothesis supported or refuted?
    e. What do you conclude about the transformation of campus?

Q40. Write your summary.

5 | Applying What You've Learned: The Human Population

1. The replacement rate for a population is 2.1 (instead of 2) to account for children who are born but do not reproduce in the next generation. Yet, populations of humans with fertility rates even lower than 2.1 can still be growing. How is this possible? Describe three ways that this might happen.

2. Think back to the excerpt from the film *Don't Panic—The Truth about Population*. Complete table 5.8 by listing the current "pin-code for the world" for each region of the world. Also list the predicted pin-code in 2050 and 2100.

Table 5.8. Pin Code for the World

Location	Pin code at present (billions of people)	Pin code at 2050 (billions of people)	Pin code at 2100 (billions of people)
The Americas			
Europe			
Africa			
Asia			

What impact will this population growth have on the biosphere if everyone in the world lives has an ecological footprint similar to the class average?

3. What is meant to ecological overshoot? What is being overshot?

5

4. What percent of the observations of campus made by the class were asphalt, concrete, or buildings? (Individually and added together.) Is this more or less than you thought it would be?

5. Using evidence from the videos, the ecological footprint data, and the World Population Data Sheets answer the following question: Is 11 billion people below, equal to, or above earth's carrying capacity for the human population? What evidence can you give to support your stance?

6 | Biogeochemical Cycles Pre-lab

1. How does an ecosystem differ from a community?

   ❑ How do biogeochemical cycles relate to an ecosystem?

2. Write out the equations for photosynthesis and respiration. What do you notice about these two processes?

   ❑ Scan the QR code or visit go.chemeketa.edu/climatechange to watch a 3-minute video from The Nature Conservancy called *A Natural Solution to Climate Change*. Why is planting trees and mangroves such a powerful part of the solutions needed to fight climate change?

❑ Where does the carbon dioxide that plants absorb come from?

3. What is pH? What is being measured by pH?

❑ Write out the numbers 0–14 O=on the pH scale in figure 6.1, then mark
which point represents a neutral pH and which direction is acidic and which
direction is basic.

Figure 6.1. pH Scale

❑ Now, place these common solutions where they belong on
the scale: pure water, rain water, sea water, coffee, lemon juice,
household bleach, battery acid, soda, household ammonia.
Use the NOAA Primer on pH found by scanning the QR code
or visiting go.chemeketa.edu/phprimer.

❑ What is NaOH? Is it an acid or a base? What is HCl? Is it an acid or a base?

4. If I say adding X to a solution will make the solution more basic, does that mean
the solution is a base? Why or why not?

Questions about items we will use in lab today:

5. What is the role of phenol red in today's experiments? What color will the phenol red turn if it is placed in an acidic solution? What color will it turn if it is placed in a basic solution?

6. From what is dry ice made? How does this molecule affect the pH of a solution?

7. Describe two safety precautions required for this lab.

6 | Biogeochemical Cycles

We have learned that communities are groups of species that interact and change over time. All communities exist in a physical environment. When we look at a community and its physical environment together, we are looking at an **ecosystem**. Our studies of ecosystems generally focus on two aspects of how the community and its environment interact, the flow of energy and the flow of matter.

All material objects, living and non-living, are made of matter and energy. Energy enters ecosystems as light, is captured by photoautotrophs and stored in the chemical bonds of organic compounds. Eventually, all energy that enters an ecosystem is radiated back into space as heat, although the energy in substances like fossil fuels and minerals may take millions of years to reach this final transition. This one way flow of energy through an ecosystem does not apply well to matter.

Matter, the substances that make up all physical objects, is used and reused by Earth's communities. Matter consists of the atoms of a variety of different elements. We refer to the paths followed by matter as it moves between living and non-living things as a **biogeochemical cycle**. It incorporates living things (the **biotic community**), **exchange pools** (environmental sources that the biotic community can access to obtain/release matter like the atmosphere and ground water) and **reservoirs** (inaccessible stores of matter like rock and fossil fuels).

Any given type of matter—we often call these nutrients—cycles both through the biotic community and the exchange pools and reservoirs. Due to this process of cycling, we can truly say that we are all built of carbon, nitrogen, sulfur, hydrogen and oxygen atoms (to name a few) that were once part of dinosaurs. In this lab you will learn about the cycling of matter by focusing on how a specific element, carbon, cycles through an ecosystem.

By the end of this lab, you should be able to:

- ❑ Draw a diagram of the carbon cycle that indicates how it cycles through the biotic community, exchange pool and reservoir.
- ❑ Conduct a test for the presence of carbon dioxide dissolved in water.
- ❑ Trace the movement of carbon through the carbon cycle.
- ❑ Describe several ways that human activity affects the carbon cycle.
- ❑ Estimate your Carbon Footprint.

Lab Safety

This lab involves the use of several hazardous chemicals. Note the following:

- **Sodium hydroxide (NaOH) and hydrochloric acid (HCl) are caustic substances.** Wear eye protection (preferably goggles) and gloves when working with these substances. If you feel burning on your hands, wash well with soap and water.
- **Phenol red** is used as a pH indicator and may cause skin irritation. In a basic solution it turns magenta (bright pink) and in an acid solution it turns yellow. Wear gloves and/or wash your hands well at the end of the lab.
- **Dry ice** (frozen carbon dioxide) is very cold. Do not handle it with your bare hands. Use the tongs provided.
- **Disposal**: All solutions in this lab can be disposed of in a sink with running water.

Exercise 6.1. The Carbon Cycle

In this lab, we will focus on the biogeochemical cycle of the element **carbon**. Life on Earth is carbon-based, which means that carbon atoms form the molecules that build the bodies of all organisms. These molecules that make up the bodies of living creatures are called **organic compounds**. Living organisms, specifically, producers, use carbon dioxide (CO_2) to build organic compounds. Carbon dioxide is called an **inorganic compound**. The energy captured by photosynthesis is stored in the chemical bonds between the carbon molecules in these compounds. This stored energy is released by consumers as they break down their food. The breakdown of food both releases CO_2 and provides simple organic compounds that can be used to build the body of the consumer. When an organism dies, decomposers break down the organic compounds and release additional CO_2 through respiration, returning the molecules to an inorganic state. Thus, carbon is cycled through the biotic community.

Since carbon is utilized in its gaseous form (CO_2), we can say that the exchange pool for carbon is the atmosphere. We will learn in this lab that carbon is also stored in ocean sediments, rocks and in fossil fuels (the carbon reservoir).

Materials

Each group of 4 students should obtain a kit containing:

- ❑ 1–100 mL beaker
- ❑ 100 mL Erlenmeyer flask with a rubber stopper and a rubber tube attached
- ❑ 2 Test tubes without stoppers
- ❑ 3 Test Tubes with stoppers
- ❑ Test tube rack

- ❑ Dropper bottle of HCl
- ❑ Dropper bottle of NaOH
- ❑ Dropper bottle of Phenol Red
- ❑ Glass stirring rod
- ❑ Test tube brush
- ❑ 1 clean straw

Other materials provided by your Instructor:

- ❑ Carboy of dH_2O
- ❑ Aluminum foil
- ❑ *Elodea*
- ❑ Fluorescent light source with test tube racks
- ❑ Dry ice
- ❑ Concentrated HCl

- ❑ Safety goggles
- ❑ Gloves
- ❑ Mortar and pestle
- ❑ Oyster shell or sand dollar test (skeleton)
- ❑ Digital balance with weighing dish

A. Animal Generation of Carbon Dioxide

When animals, like us, consume organic molecules in our food, we break them down to release energy for our own use. So, what happens to the carbon that those molecules are made from?

Procedure

1. Add 50 mL of deionized water and add 20 drops of phenol red indicator to the 100-mL beaker.
2. Stir with the glass rod to mix.

> **Note:** What color is your solution? If it is not magenta (bright pink), add a few drops of NaOH to make it an orangy-red or light salmon color.

3. One member of your group should blow through the straw into the solution until you observe a color change.
4. Dispose of your solution in a sink with running water.

Q1. What color does the solution turn?

Q2. What do you think was the cause of this color change?

Q3. What reaction do you think is taking place to make the solution change color? (Think back to the pre-lab about how pH is measured.)

> Skip ahead to Part E. Part E requires 2 hours to run and we want to make sure to get it set up.

B. Performing a Titration

The **pH scale** provides a tool to measure the acidity of a solution. Low pH numbers indicate that a solution is **acidic** while high pH numbers indicate that it is **basic**. PH 7 is considered **neutral** pH (neither acidic or basic). In much of this lab, we will be looking for a color change in an indicator chemical due to a shift in the pH of the solution. An acidic solution will have a different color than a basic solution. The color change occurs with a very small change in pH so we want to adjust the initial color so that it is very near to the "tipping point" where we will detect that small change in pH. The first steps direct you in slowly adjusting the initial pH by adding a weak acid, or base, one drop at a time (**titrating**) to achieve the desired initial color. A small change in subsequent pH of the solution will be easily detected.

Procedure
1. Add 50 mL of deionized water to the 100 mL beaker.
2. Add 20 drops of phenol red and stir with the glass-stirring rod.
3. **Put on your goggles.**
    a. If your solution is yellow, it is acidic. Add NaOH (sodium hydroxide) one drop at a time, stirring after each addition until the color just starts to turn pinkish-orange.
    b. If your solution is magenta, it is basic. Add HCl (hydrogen chloride) one drop at a time, stirring after each addition until the color just starts to turn pinkish-orange.
4. Repeat this process several times changing the color back and forth for practice.

Q4. How many drops are required to change your solution from magenta to yellow?_____

Q5. How many drops are required to change your solution from yellow to magenta?_____

You have just performed a titration. The benefit of having your solution at its tipping point is that very small changes in pH will be easy to detect. You will be asked to make titrated solutions a few more times in this lab. For now, you can **use this titrated solution to do part C**.

C. What is the Cause of the pH Change?

When we blew into the phenol red solution, the solution eventually became yellow. Why? We have discussed the fact that animals release CO_2 as a result of the breakdown of the organic compounds they obtain from their food, but there are other molecules present in the air we exhale, such as nitrogen and even oxygen. So is it the carbon dioxide that causes this color change?

Carbon dioxide is generally available to us as a gas, but we can also obtain it in the solid form we know as dry ice. Dry ice is quickly converted from solid to gas (skipping the liquid stage) at room temperature. This means that if we let the dry ice melt, we should get CO_2 gas. We can "bubble" that gas into the phenol red solution to determine if CO_2 actually causes a pH change.

Procedure
1. Prepare 50 mL of titrated Phenol Red solution (Use solution from part B).
2. Divide the solution between two clean test tubes. Make sure both are the same pink-orange color. (Magenta is also okay.)
3. Place the two tubes on opposite ends of a test tube rack.
4. Put the end of the rubber tubing into the solution in one test tube, leaving the stopper end free.
5. Add approximately 50 mL of tap water to your Erlenmeyer flask.
6. Obtain a small piece of dry ice from the instructor and add it to the water in the Erlenmeyer flask. The water will cause the carbon dioxide to vaporize more rapidly.
7. Rapidly put the stopper on the Erlenmeyer flask to direct carbon dioxide through the rubber tubing into the phenol red solution in one of the test tubes.

Q6. What happens to the phenol red solution?

Q7. What happened in the second test tube during the same time period?

Q8. What is the role of the test tube that does not have CO_2 bubbled into it in this experiment?

Q9. What is the most likely cause of the color change in this experiment?

8. Empty the contents of the flask and the tubes into a sink with running water.

Understanding What You Observed

When carbon dioxide dissolves in water, some of it reacts to form small amounts of carbonic acid. As an acid, it reduces the pH of the solution. Changes in environmental pH can cause the compounds dissolved in a solution to change shape. In the case of pigmented molecules like phenol red, this change in shape causes the molecule to reflect light differently, resulting in the color change that we see. The formation of carbonic acid and its interaction with calcium is critical to the formation of the shells of many ocean creatures. We will revisit this concept below.

D. Animal Storage of Carbon Dioxide

Like plants, animals also use some of the carbon they consume to build their bodies. The incorporation of carbon into the tissues of the body is key to the biotic community and to the formation of a carbon reservoir. This experiment will demonstrate that carbon is stored in the bodies of animals. The piece of shell that you will use comes (most likely) from a sand dollar. Sand dollars are a type of sea urchin in the phylum Echinodermata. They live in soft bottom sandy areas and eat crustacean larvae, copepods, diatoms, algae, and detritus. In this exercise, they serve as just one example of an animal with carbon in its tissues.

Procedure
1. Prepare 50 mL of titrated Phenol Red solution (see part B).
2. Divide the solution between two clean test tubes. Make sure both are the same pink-orange color. (Magenta is also okay.)
3. Place the two tubes on opposite ends of a test tube rack.
4. Weigh out 1 g of shell.
5. Grind the shell to a fine powder with the mortar and pestle and pour the powder into the empty Erlenmeyer flask.

6. Put the end of the rubber tubing into the solution in the test tube leaving the stopper end free.
7. Ask your instructor to add 20 drops of concentrated acid to the powdered shell
8. Quickly insert the stopper into the Erlenmeyer flask and swirl it.
9. Allow the gas generated to bubble through the phenol red solution.

Q10. What happens to the phenol red solution?

Q11. What happened in the second test tube during the same time period?

Q12. Based on these observations, what gas is produced by the shell when treated with concentrated acid?

Q13. How did this substance get into the animal shell/skeleton?

Q14. What parts of your body might also produce this substance if exposed to concentrated acid?

Q15. Do animals store carbon in the tissues of their bodies?

10. Empty the contents of the flask and the tubes into a sink with running water.

Why Is This Important?

We have just demonstrated that animals incorporate carbon into their bodies. This may not seem like a big deal, but it is. The shells we used for this experiment come from marine animals.

The ocean is full of life. All of these organisms incorporate carbon into their tissues. Many of them use the carbon to make shells and other structures. Ultimately, the organism dies and because it has mass, sinks to the bottom of the ocean forming **ocean sediments**. As ocean sediments accumulate, pressure builds up and lower layers of sediment are converted to rock (including limestone and marble). Ocean sediments make up a portion of the **reservoir** where carbon is trapped and inaccessible to the biotic community.

Q16. Rain and groundwater are slightly acidic in pH. How do you think that exposure of limestone and marble to the atmosphere affects the carbon cycle?

Q17. Another effect of increased carbon dioxide in the atmosphere is an increase in the absorption of carbon dioxide into our oceans. This makes ocean water slightly more acidic than it used to be (though it still has a pH > 7). What effect do you think a slightly more acidic ocean has on marine organisms that make shells? Why?

E. Plants and Carbon Dioxide

Terrestrial plants have access to gaseous CO_2 in the atmosphere and aquatic plants are in contact with it dissolved in water. We have discussed the fact that plants use CO_2 when they perform photosynthesis. Let's do an experiment to confirm our hypothesis. Do aquatic plants use carbon dioxide in the water around them?

Procedure

1. Add 75 mL of deionized water into a beaker.
2. Add 30 drops of phenol red indicator solution and stir with the glass rod to mix (if your solution is not red, please consult with your instructor).
3. Use a clean straw to blow into the beaker. Blow slowly and keep a careful eye on the color. As soon as the solution turns yellow, stop blowing!
4. Fill the three test tubes (the ones the stoppers fit in) with the solution and double check to ensure that they are each the same color.
5. Obtain two fresh sprigs of healthy *Elodea* the same length as the depth of solution in your test tubes. Rinse the *Elodea* with deionized water and insert the sprigs cut end down into two of the test tubes until fully submerged.
6. Place the stoppers on all three test tubes.
7. Wrap one of the tubes containing Elodea in foil so that no light can enter the tube.
8. Place all three tubes in the tin baking pan below the intense fluorescent light source for 2 hours.

Answer questions 18–22 before moving on

Q18. Why did we blow into the Phenol red solution before adding it to the test tubes?

Q19. What would you expect to happen to the color of the solution if the plant produces CO_2? (That is if the plant releases CO_2 into the solution.)

Q20. What would you expect to happen to the color of the solution if the plant uses CO_2? (That is if the plant absorbs CO_2 out of the solution.)

Q21. Why did we make a tube that had phenol red solution and no plant?

Q22. Why did we make a tube that had phenol red solution and a plant and then wrap it in foil?

At the end of two hours, answer questions 23–27.

Q23. What happened to the color of the solution in the tube with the plant?

Q24. What happened to the color of the solution in the tube that contained the plant and was wrapped in foil?

Q25. What happened to the color of the solution in the tube without the plant?

Q26. Based on your observations, what does a plant do with carbon dioxide when it is exposed to light?

Q27. Based on your observations, what does a plant do with carbon dioxide when it is kept in the dark?

9. Place the plants in the recovery beaker then empty the contents of the flask and the tubes into a sink with running water.

Exercise 6.2. Examining the Carbon Cycle in the Ocean

As you have seen in Exercise 6.1, carbon moves through producers and consumers and back to the atmosphere through the carbon-containing compounds glucose and carbon dioxide. In Exercise 6.2 you will begin to get a sense of how carbon moves through ocean systems. To begin, carbon dioxide in the atmosphere interacts with water in the oceans and, as you saw with our straw experiment in which you blew air into water, makes the water more acidic. Every year, the oceans absorb about 25% of the CO_2 in the atmosphere. In 1865 CO_2 levels in the atmosphere were measured at 275 ppm (parts per million) and by 2020 that amount had increased significantly to 440 ppm. Because carbon dioxide in the atmosphere is at historic high levels, our oceans are becoming more acidic (or less basic) at an alarming rate. Ocean acidification (or OA) affects the life cycle of organisms like the sand dollar that you examined in Exercise 6.1D and the formation of coral reefs. OA also affects phytoplankton that live on the surface of the ocean and limits their collective ability to produce oxygen.

A. Ocean Acidification in a Jar*

*Adapted from this website: https://www.exploratorium.edu/snacks/ocean-acidification-in-cup

In this exercise, you will create a microcosm of the ocean interacting with the atmosphere. Using a different pH indicator than we used in Exercise 6.1, bromothymol blue (because it is blue, like the ocean), you will observe how adding carbon dioxide to the atmosphere changes the pH of the water.

Materials
- ❏ Safety goggles
- ❏ An acid-base indicator such as bromothymol blue, diluted with water: 8 milliliters bromothymol blue (0.04% aqueous) to 1 liter of water
- ❏ Two clear pint glass jars
- ❏ One small (3 oz) paper cup
- ❏ Masking tape
- ❏ Plain white paper
- ❏ Permanent marker
- ❏ Baking soda
- ❏ White vinegar
- ❏ Two Petri dishes to use as lids for the plastic cups
- ❏ Measuring spoons

Procedure

1. Put on your safety goggles.
2. Pour 40–50 mL of acid-base indicator solution into each of the two glass jars.
3. Add 1/2 teaspoon of baking soda to the paper cup.
4. Tape the paper cup inside one of the clear plastic cups containing the indicator solution so that the top of the paper cup is about 1/2 inch (roughly 1 centimeter) below the top of the glass jar. (There needs to be room between the top of the paper cup and the lid for the jar.). Make sure the bottom of the paper cup is not touching the surface of the liquid in the jar—you don't want the paper cup to get wet. The second glass jar containing indicator solution will be your control.
5. Place both jars onto a sheet of white paper and arrange another piece of white paper behind the jars as a backdrop (this makes it easier to see the change).
6. Carefully add 1 teaspoon of white vinegar to the paper cup containing the baking soda (image below). Be very careful not to spill any vinegar into the indicator solution. Immediately place a Petri dish over the top of each plastic cup.

Q28. What happens to the baking soda when you add the vinegar? What gas is being released into your atmosphere?

Q29. What change do you see taking place in your ocean? Where is this change happening? Why is this change happening? Is your ocean more acidic or more basic than your control?

Q30. After a few minutes observe your ocean again. What changes do you see now? Why did these changes continue to occur?

Q31. Relate your ocean-atmosphere microcosm to the real ocean and atmosphere. What change have you just modeled? What is the name of the process you just observed? In the real ocean does this process make the ocean water an acid?

B. Exploring the Virtual Urchin – Our Acidifying Ocean Site

In the real ocean real creatures are negatively affected by ocean acidification. To understand the consequences of OA in greater detail, we will explore the website Our Acidifying Ocean – the Virtual Urchin tab by tab. Answer the questions below as you move through the tabs:
Title | Intro | Air pH | Ocean pH | Chemistry | Levels | Diversity | Cycles
(Note: we will not be doing the virtual lab outlined in the "How to" tab.)

Q32. In the Air tab, what is the difference between the wavy blue line and the purple line? What was the concentration of carbon dioxide in the atmosphere (roughly) in 1960? For the year you were born? In 2020?

Q33. From the pH tab: redraw the pH scale you made below.

❑ How acidic is ocean water normally? Is it actually an acid? Why do you say this?

❑ What sort of scale is the pH scale?

❑ So how much more acidic is a pH of 5 compared to a pH of 6? How much more acidic is a pH of 5 compared to a pH of 7?

❑ What can you hypothesize, therefore, about small changes in the pH scale number for an organism trying to live in this changing environment?

Q34. From the Ocean pH tab:

❑ Describe how ocean pH in the oceans has changed over the last 20 million years.

❑ Describe how ocean pH has changed since the industrial revolution.

❑ Describe the projection for the change in ocean pH into the future. In the year 2100 will the ocean actually be an acid?

❑ So what do we mean by ocean acidification?

Q35. From the Chemistry tab: In a few comparative sentences using the words carbon dioxide, bicarbonate, etc (instead of the chemical formulas), interpret the equations that describe how carbon dioxide interacts with water:

❏ Under normal conditions

❏ Under acidified conditions

Q36. From the Levels tab:

❏ What is the pH for each level? (Optimistic, middle ground, pessimistic?) Where do these scenarios come from?

❏ How will ocean creatures that build body structures out of calcium carbonate be affected by acidified conditions?

❏ Why will they be affected this way?

Q37. From the Diversity tab: Which ocean creatures will be affected and which won't be?

Creatures affected by OA | Creatures NOT affected by OA

Q38. From the Cycles tab: When during their life cycle are the calcifiers like sea urchins most affected? Click through the life cycle and watch the video to answer this question.

Q39. Summarize what you have learned about the carbon cycle in the oceans by describing three facts or take-home messages.

NOTE: We are not actually going to do the virtual lab in which you examine sea urchin larvae but feel free to continue exploring as your time allows. No need to click on the How to tab.

Exercise 6.3. Modeling the Carbon Cycle (optional)

Using the laminated cards with terms from the carbon cycle, work with a partner to "build" a model of the cycle. Use the arrow cards to indicate the movement of carbon. Include the ways in which humans influence the cycle and the consequences of human actions. These consequences include global warming and ocean acidification. Once you have put all your laminated cards in position, sketch the cycle below.

6 | Applying What You've Learned: Biogeochemical Cycles

1. In the space below draw the **biotic cycle** (producers → consumers → decomposers). On the cycle, indicate what form carbon is in (CO_2 or sugars) in each of the parts of the biotic cycle, and what process is used to transfer the carbon between each part of the cycle (consumption, death and decay, photosynthesis, or respiration.)

2. Using the figure you just drew above, explain from where and in what process do animals get the carbon they need to build organic molecules, such as proteins and fats. How did Exercise 6.1.D illustrate this process (Animal Storage of Carbon Dioxide)? Be sure to indicate what happened in the experiment and how your control helped verify the result.

3. Using the figure you just drew above, explain from where and in what process do plants get the carbon they need. How did Exercise 6.1.E (Plants and Carbon Dioxide) illustrate this process? Be sure to indicate what happened in the experiment and how your controls helped verify the result.

4. Compare the carbon cycle on land with the carbon cycle in the ocean. How are they similar? How are they different?

5. What is the cause of ocean acidfication? Outline the process by which ocean acidification affects calcifying species?

7 | Bacteria and Protist Diversity Pre-Lab

Use this week's lab packet, your textbook, or other resources to answer the following questions.

1. Create a timeline from the information provided in the introduction to the lab that lists the year, the scientist, and a few words about their discovery. You should have at least 6 dates/scientists. What can you generalize about scientific discovery from this brief history?

2. Watch the video, Seeing the Invisible, about van Leeuwenhoek's exploration of the microscopic world. List 5 things that van Leeuwenhoek was the first person to see. What would you do if you were the first person in the world to see or discover something? Scan the QR code or visit go.chemeketa.edu/animatedlife to watch a video about van Leeuwenhoek's first glimpse into the microscopic world called "Seeing the Invisible."

3. What is a virus? Sketch and label a diagram of the parts of an enveloped virus in-cluding the genome (could be DNA or RNA in different forms), the capsid (protein coat), the envelope, and the glycoproteins (spikes).

4. What is a bacteria? Sketch and label a diagram of a bacteria including the DNA, the ribosomes, the plasmid, the cell membrane, the cell wall, the capsule and pili, and the flagella.

7

5. What is a protist? Sketch and label a diagram of a protist. There are many types, so you may choose a diatom, a paramecium, or a spirogyra. There are many oth-ers but these three we are likely to see in lab.

7 | Bacteria and Protist Diversity

Viruses, bacteria, and many protists are microscopic and because of this, they are not as well understood as the macroscopic plants, fungi, and animals that we will be learning about in the coming weeks. Scientists have been puzzling over the evolutionary history, roles in the environment, life histories, and relationships to humans of these small but plentiful creatures for about 350 years. In 1665, Robert Hooke looked through a microscope at cork and noted that the shapes he saw looked like the cellula that monks lived in at the time and called these shapes **cells**. The cork cells Hooke saw were the remains of the cork tree's cell walls, and so were not living. They showed none of the internal compartments of cells that would later be discovered.

In 1674 Antonie van Leeuwenhoek peered through one of his microscopes made of specially ground glass to look at pond water and saw the green algae, **Spirogyra** (a protist), and living bacteria for the first time. He called the organisms he saw in his drop of pond water "animal-cules" or little animals. The rest of the scientific world was skeptical of his report, but when other scientists repeated the studies, a general acceptance of microscopic cells being the basis of all life gained support. Eventually the first cell theory was formulated in 1838 by Theodore Schwann and Matthias Jakob Schleiden.

The discovery of **viruses**, which are much smaller than bacteria, would take another 50 years. In 1892 Dmitri Ivanovsky was studying an illness in tobacco plants and put extracts of diseased leaves through a filter that had pores small enough to prevent bacteria from getting through. He tested the filtration and found it could still infect tobacco plants with the illness.

Six years later, in 1898 Martinus Willem Beijerinck called the infective agent a "virus" and the study of the tobacco mosaic virus has been a central part of virology ever since. After this, many other viruses were studied and named, most of which were diseases of poultry and livestock. These viruses had to be characterized by the symptoms they caused because they were too small to be seen with the microscopes of the time. Wendell Meredith Stanley suggested that viruses were discrete particles rather than liquid, but it wasn't until 1930's that the first virus was "seen" through an SEM.

The designation of the **protists** into subgroups also took a number of decades to resolve from their first documentation by van Leeuwenhoek. After considerable debate on the nature of the small and microscopic life forms that were found in pond water (among other locations), TL Jahn united the algae, protozoa, and some other groups into the Protista in the late 1940s.

This kingdom is still recognized today, despite the fact that it is paraphyletic, which means that the organisms that are classified as Protista do not share a set of unique distinguishing features that makes them similar to each other and different from other groups. Instead, protists are definable only as eukaryotes that are not plants, animals and fungi.

Biological **communities** differ in terms of their composition and diversity. The composition of a community is merely a list of species found in that community. The diversity of a community is based on the number of different species present (species richness) and the relative abundance (percent of the total number of all individuals) of each species. In everyday language, people use the word "diversity" when really they mean "species richness," because they are really just referring to the number of species present. Today we will mostly be focusing on species richness.

By the end of this lab, you should be able to:

- ❑ Identify the main characteristics of bacteria and protists.
- ❑ Understand antibiotic resistence and measure the effectiveness of different antibiotics on two common bacteria.
- ❑ Explore the human microbiome.
- ❑ Examine and identify protist diversity in a drop of pond water.
- ❑ Create a pie chart using pond diversity data.

Exercise 7.1. Bacteria

Bacteria are all around us. When we see evidence of it, like in figure 7.1, we are often grossed out.

Figure 7.1. Bacterial Colonies on a Petri Dish

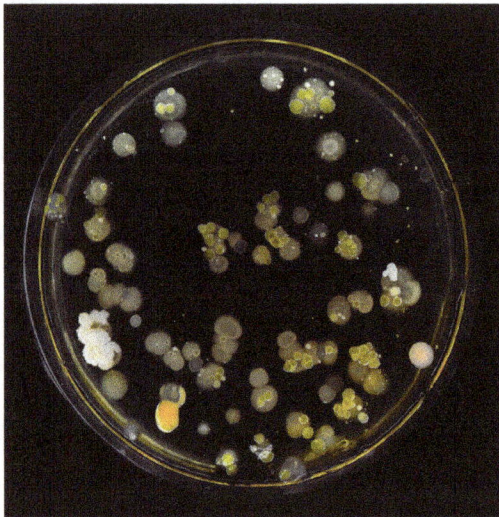

And yet, probably all of what you see on the petri dish is harmless to humans. But there are some bacteria, like MRSA (Methicillin-Resistant *Staphylococcus aureus*), that have developed antibiotic resistance, which renders the medicines we have to fight the bacteria useless. We are going to explore the nature of antibiotic resistance and the importance of the gut biome to human health, but during this lab we will return to the observational study we started last week to document the diversity of (mostly) bacteria in our classroom environment.

A. Understanding Antibiotic Resistant Bacteria

To learn more about what MRSA is and how it became resistant to antibiotics, scan the QR code or visit go.chemeketa.edu/goodbacteria, watch the video, and answer the questions.

"Science Bulletins: MRSA—When Good Bacteria Go Bad" by the American Museum of Natural History

Here's the blurb about this video: The bacteria Staphylococcus aureus is a benign resident of the human microbiome. But in the last 15 years, a strain of it has evolved to become a major public health problem. To understand how everyday microbes can change so dramatically, laboratories are investigating how bacterial communities exchange genetic information in the environments we share.

- ❑ How do babies acquire their microbiome (the bacteria that live in and on their bodies)?
- ❑ When did MRSA show up in hospitals? When did it show up in communities outside of hospitals?
- ❑ What is SPE-G? What is the benefit to MRSA to have this gene?
- ❑ How do humans give off bacteria?
- ❑ Because most of the bacteria that inhabit us are harmless, what is the disadvantage of using antibacterial soaps and hand sanitizers?

With this understanding of antibiotic resistance, examine the images provided for you of two petri plates. The one on the left was plated with Eschericia coli and the one on the right was plated with Staphlococcus aureus. Once growing on the petri dish, each type of bacteria was then subjected to different antibiotics (saturated on the paper discs). Around each disc you can see a "zone of inhibition" where the bacteria would not grow (or were inhibited by the antibiotic.) Using the ruler and scales provided, answer the following questions:

Q1. How large is the zone of inhibition for each of the antibiotics:

Table 7.1

	E. coli	S. aureus
AM		
S		
TE		
C		
P		
N		

Q2. Which antibiotic was most effective against E. coli? Which was the least effective?

Q3. Which antibiotic was most effective against S. aureus? Which was the least effective?

Q4. Use the chart provided under the pictures of the petri dishes to determine the answer to the following questions.

❑ Which antibiotics is E. coli resistant to? Which is it sensitive to?

❑ Which antibiotics is S. aureus resistant to? Which is it sensitive to?

Q5. Create a graph or an infographic to convey the information you learned above to an 8th grader on a separate piece of paper.

B. Understanding the Bacteria Associated with Humans

So if MRSA is one bacteria you hope not to encounter during your lifetime, what are some of the bacteria you encounter regularly? Scan the QR code or visit go.chemeketa.edu/discovermagazine to see a slide show of bacteria that make up your microbiome created by authors at *Discover Magazine*. Answer the questions below. (There is one question for each picture.)

Q6. Describe the different types of "habitats" bacteria might find on your skin in different parts of your body.

Q7. How many bacterial species reside in a typical gut biome? How are these bacteria important to your health?

Q8. Compare and contrast the skin and gut bacteria of babies born vaginally with that of babies born by C-section.

Q9. How does the gut biome change as a baby gets older? What does the child gain from its gut bacteria?

Q10. What is one consequence of eating processed food on your gut biome?

Q11. How can the bacteria living in our gut make us fat? What is the main difference between the gut bacteria of fat humans and thin humans?

Q12. Describe two ways that our gut bacteria can influence actions we take that seem unrelated to eating.

Q13. What can the gut biome of different species (such as gorillas) tell us about their evolution?

Q14. How do some viruses use our own gut bacterial to infect us?

Exercise 7.2. Protists (and Other Organisms) in a Drop of Water

Aquatic communities contain microscopic organisms from all the domains of life. In this exercise, we will focus on the microscopic organisms known as **plankton** that passively drift with water currents within the water column or swim weakly. The variety of plankton in an aquatic community can be divided into producers, consumers, and decomposers. The **producers** usually obtain energy through photosynthesis and have a bright color, while the **consumers** obtain their energy from eating other organisms and the **decomposers** from breaking down the dead remains of both producers and consumers.

Most producers in aquatic ecosystems belong to the kingdom Protista (protists). They are commonly called **algae** (singular is "alga"). The next most common producers in freshwater ecosystems are the **cyanobacteria**, which, as you would expect, belong to the kingdom Bacteria.

Most producers found in aquatic communities are lumped together as **phytoplankton**. The consumers that live in the plankton are in the kingdoms Protista and Animalia and are collectively known as **zooplankton**. While you will see zooplankton in your samples, they are much less common than phytoplankton. All of the common freshwater decomposers are either bacteria or the fungal-like protists known as water molds, thus are either too small to be seen with our microscopes or too rare. As a consequence, they will not be included in today's exercise.

A. Making a Protist Guide While Learning to Use a Microscope

Recall from the introduction to this lab that protists are grouped together because of what they are not, rather than what they are. Remember that a protists is a eukaryotic organism that is not a plant or animal or fungus. Many, but not all, protists are single-celled organisms. Some are colonial. A few are multicellular, like seaweeds and slime molds. Though these macroscopic protists are quite interesting, they are not the focus of today's lab exercise. Instead we will focus on protists that you might find in a drop of pond water. To see these protists you, like van Leeuwenhoek, will need to use a microscope.

Compound Microscope Introduction & Making a Wet Mount
If this is the first time you have used a microscope, refer to the directions by your instructor. You may have had a virtual microscope assignment to do before the lab or you may get an introduction from your instructor. If you are doing the introduction in lab, see the "Compound Microscope Introduction and Making a Wet Mount" instructions provided in the appendix of this lab manual.

Procedure
1. Create your own guide to identify protists in pond water. *Note: Your instructor may decide to do this procedure in reverse order and have you start with #4 to explore a sample of actual pond water, before you begin to look at the video and virtual pond water as guides. You might also just consult the paper guides available in the lab.*
2. Examine the video clip either with your class or on your own to identify the types of organisms you will see in the pond water samples in Exercise B. https://www.youtube.com/watch?v=PPls7CQxOBA
3. As you see each protist, make observations about its behavior in table 7.3 and make a quick sketch that will help you recognize it again when you see it in a pond water sample. You might jot down identifying characteristics, like "is round and green" or "long and stringy." You may refer to the "Virtual Pond Dip" (http://www.microscopy-uk.org.uk/ponddip/) diagram or other references provided for each group for some of the organisms in the boxes as needed.

Table 7.3. Guide to Common Protists Seen in Pond Water

Amoeba (Amoeboid Zooplankton)	Vorticello (Ciliate Zooplankton)
List behavior, characteristic, and sketch	List behavior, characteristic, and sketch
Stentor (Ciliate Zooplankton)	**Paramecium (Ciliate Zooplankton)**
List behavior, characteristic, and sketch	List behavior, characteristic, and sketch
Euglena (Flagellate Zooplankton)	**Volvox (Motile Phytoplankton)**
List behavior, characteristic, and sketch	List behavior, characteristic, and sketch
Diatoms (Non-motile Phytoplankton)	**Spirogyra (Non-motile Phytoplankton)**
List behavior, characteristic, and sketch	List behavior, characteristic, and sketch

Q15. Which of the protists you sketched are photosynthetic? (You would guess they are photosynthetic if they are green, but some protists are photosynthetic and have more of a brown or golden pigment.)

Q16. Which of the protists you sketched have flagella (one or two long whip-like tails)?

Q17. Which of the protists you sketched have cilia (lots of short hairs around their edges)?

Q18. Does the amoeba have any of the properties mentioned above? Describe a characteristic that sets it apart from the other protists.

In creating your guide and answering the questions above, you have begun to learn how scientists classify protists. Phytoplankton are usually blue-green, golden, or grass green in color while zooplankton tend to be unpigmented except, sometimes, they take on the color of the food they consume. Plankton are further classified *by the way that they move*. Some types of plankton, including many types of green algae and diatoms, are non-motile, meaning they don't move on their own but can drift with the water currents—even on a microscope slide.

Other types of green algae, euglenoids and many types of zooplankton have one or more long whip-like extensions (flagella). These organisms, known as **flagellates**, move by undulating their flagella. Many types of zooplankton have many short, hair-like projections (cilia). **Ciliates** move by coordinated, oar-like flapping of their cilia. A final group of protists, known as **amoeboids**, move by extending a foot-like projection (pseudopod) made from their plasma membrane. The rest of the cell flows into the pseudopod moving the organism forward. And now it is time to search for these organisms (and others) in a drop of pond water.

B. Exploring the Species Richness in a Drop of Pond Water

Procedure

1. Obtain the water samples from one of the locations.
2. Make a wet mount from your sample.
3. Examine your slide with your compound microscope. Scan back and forth across the slide in a systematic fashion.
4. Also examine the slide at each level of magnification so that you have the opportunity to see organisms of various sizes.
5. Classify each different kind of organism you find into one of the categories listed in table 7.4 and fill in the table as you go.

a. Give any group you see any representative of in your sample a tally mark for each individual you see. For organisms like spirogyra, make your best estimate.
b. Note that animal species that you are likely to see have been added to the table. Refer to the Lab Bench Information and/or guides provided to see what each of these organisms looks like. And while it is exciting to see these tiny creatures, and even worthy of a shout of "Eureka!", please remember that they are not protists.

Table 7.4. Protist (and other organisms) Abundance by Type Found in Pond Water

Type of Organism	Number of organisms seen	Percent of total
Non-motile phytoplankton: (Protista: diatoms, *Spirogyra*; Bacteria: cyanobacteria)		
Motile phytoplankton: (Protista: motile green algae, *Volvox*, *Chlamydomonas*)		
Flagellate zooplankton: (Protista: *Euglena*)		
Ciliate zooplankton: (Protista: *Stentor*, *Vorticello*, *Paramecium*)		
Amoeboid zooplankton: (Protista)		
Rotifers: (Animalia)		
Crustaceans: (Animalia: ostracods, copepods, water fleas)		
Worms: (Animalia)		
Insects: (Animalia)		
Total number of organisms seen		(Divide # of each type by total)

Q19. Which types of species were most abundant in your sample? Which was least abundant?

Q20. Do you think looking at one slide from your sample gives an adequate representation of the organism present in that pond? Why or why not?

Q21. Compare your data with the data from your partner. Did you find the same types of species in the same abundances? Why do you think this is the case?

Q22. Think about the source of each of the samples. What factors might account for the differences in species richness found in different ponds?

6. Create a graph of your data using the pie chart outline provided. Group the different species of protists into these broader categories to maximize the number of individuals in each category and make fewer wedges in your pie chart.

    Graph Title_____

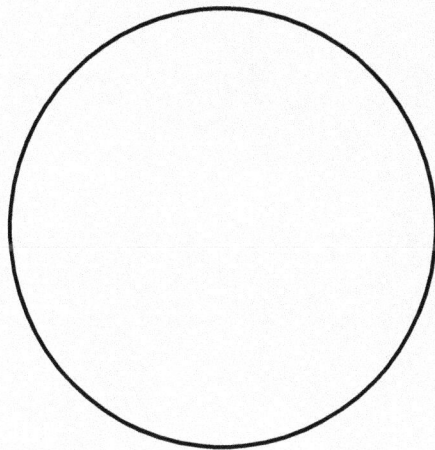

7 | Applying What You've Learned: Bacteria and Protist Diversity

1. What type of organism (virus, bacterium, or protist) can be treated with an antibiotic medicine? What does it mean when we say an organism of this type is "antibiotic resistant?" Why are we as a society concerned about antibiotic resistance?

2. Your microbiome is with you to stay, whether you like it or not. Of all the aspects of the microbiome you learned about today, which one was most interesting? Why? What can you do to promote a healthy microbiome?

3. Give three examples of specific organisms that you saw in this lab and indicate whether they are producers (can do photosynthesis) or consumers (cannot generate their own food source).

4. What characteristics do diatoms and a filamentous green algae like Spirogyra have in common? What characteristics make them different?

5. Which types of organisms were most abundant in your pond water sample? Which types of organisms were the least abundant? Why do you think this was the case? Relate this information to aquatic food webs.

6. What was the most interesting thing you learned during this lab?

8 | Animal Diversity Pre-lab

We would like to be able to take you on a grand field trip to all of the exotic locations where the rich diversity of animal life can be found. Since we do not have the time or budget to do this, we have assembled a representative of pictures of specimen for you to view. Looking back over the term and the organisms we have been investigating, including the plants, fungi, protists, and bacteria, let us now consider the animals.

1. What specific traits characterize all animals? How can you distinguish animals from other forms of life? List 3–4 traits. All animals have…

    ❏ _____
    ❏ _____
    ❏ _____
    ❏ _____

2. Define the following terms:

    ❏ Primitive trait:

    ❏ Derived trait:

    ❏ Phylogeny:

3. It can be useful to know the relative phylogenetic age (which groups are geologically older than others) of different groups. What might help you to determine this relative age?

4. Explain the term **convergent evolution** and give an example. Are two species that have converged on similar traits evolutionarily closely related? For example, do you think birds and butterflies are close together on a phylogeny because they both have wings?

5. There are approximately two million named, described species. Without researching the answers, rank the taxonomic groups in table 8.1 in terms of abundance (number of different species) and age. If you were a taxonomist, which traits would you use to organize these groups within the Kingdom Animalia? What traits are unique to each group and can be used to tell each group apart?

Table 8.1.

Taxonomic group	Which group has the most species in it? Which has the least? (Rank them 1 = most, 10 = least)	Phylogenetic age (1 = oldest [been on earth the longest], 10 = newest)	Shared derived traits that this group possesses (what characteristic does this group have that differ from all other groups?)
Amphibians (frogs and salamanders)			
Birds			
Cnidarians (sea anemones and sea jellies)			
Fish			
Insects			
Mammals			
Mollusks (clams, octopus, etc)			
Reptiles (lizards, snakes)			
Sponges			
Worms			

8 Animal Diversity

Animals are multicellular, ingestive, eukaryotic, heterotrophs that diverged from their common ancestor with other eukaryotes about 700 million years ago. Additionally, most animals are mobile for at least some part of their life cycle, even though some adult forms are sessile. There are 36 animal phyla but the vast majority of species are members of the 9 main phyla:

- ❑ Porifera (sponges)
- ❑ Cnidaria (jellies, corals, anenomes)
- ❑ Platyhelminthes (flatworms)
- ❑ Annelida (segmented worms)
- ❑ Mollusca (gastropods, bivalves, cephalopods, chitons)
- ❑ Nematoda (roundworms)
- ❑ Arthropoda (arachnids, myriapods, crustaceans, insects)
- ❑ Echinodermata (sea stars, urchins)
- ❑ Chordata (some invertebrates, fish, amphibians, reptiles, birds, mammals)

In today's lab you will get to explore five of these phyla in greater detail in Exercise 8.1 by watching videos selected from a website called *Shape of Life*. You will then, in Exercise 8.2, get to explore how to make a **phylogeny**, or evolutionary tree, that outlines the relatedness between some representatives of these major phyla. And finally, in Exercise 8.3 you will make your own observations of various traits organisms in these nine phyla may or may not have in common and then you will sort out your own phylogeny for ten of the organisms you observe.

By the end of this lab, you should be able to:

- ❑ Understand the characteristics that define the five of the nine main phyla, making each unique.
- ❑ Construct a phylogeny based on a set of characteristics that are shared among a set of organisms.
- ❑ Compare some of the major features of extant animal groups including: sponges, cnidarians, worms, mollusks, arthropods, echinoderms, and vertebrates.
- ❑ Discern relationships between animal groups and draw logical conclusions about which groups are closely related and which groups are distantly related.
- ❑ Appreciate the diversity of life in the Kingdom Animalia and its value.

Exercise 8.1. Building a Phylogeny*

To understand phylogenetic trees, you need to understand something called common ancestry. You likely know that chimpanzees are the closest living relatives to humans. What that means is that, if you go back far enough— more than seven million years!—you'd find a population of animals that had some traits in common with just humans, some traits in common with just chimps, and probably a lot of traits in common with both humans and chimps.

This population of animals is the common ancestor of humans and chimpanzees. If you could follow this population through time, at some point something would happen that would separate the population into two different groups. Over time, one of these groups would become more and more chimp-like, and the other group would become more and more human-like.

By comparing traits that an organism has in common with some groups of species but is different from other groups of species, a scientist can trace which species are more closely related and which species are more distantly related. In Exercise 2, you will use this reasoning to find evidence that can determine patterns of relatedness among different animal groups.

Procedure

1. Table 8.2 includes important characteristics of six different groups of animals. Study the table to see which groups share the most traits in common and which groups have the fewest traits in common.

Table 8.2. Some Characteristics of Animal Groups

	Has specialized cells with nuclei	Has tissue	Has organs	Has a head	Has jointed arms and legs
Beetles	Yes	Yes	Yes	Yes	Yes
Crabs	Yes	Yes	Yes	Yes	Yes
Jellyfish	Yes	Yes	No	No	No
Snails	Yes	Yes	Yes	Yes	No
Sponges	Yes	No	No	No	No
Seastars	Yes	Yes	Yes	No	No

---

* (Adapted from "Pipe Cleaner Model of Animal Evolution" by *Shape of Life*.

Q16. Which groups of animals have the most traits in common?

Q17. Which groups of animals have the fewest traits in common?

Evolutionary trees model relationships among taxa. Every taxon has a unique lineage, or evolutionary history. You can think of a lineage as the path back through time from the present to the beginning of all life. If you trace back any two lineages through time, eventually they will join up. On a tree, lineages join at nodes. A node represents the time when one lineage split off from the other lineages on the tree, as seen in figure 8.1.

Figure 8.1. An Insect Tree

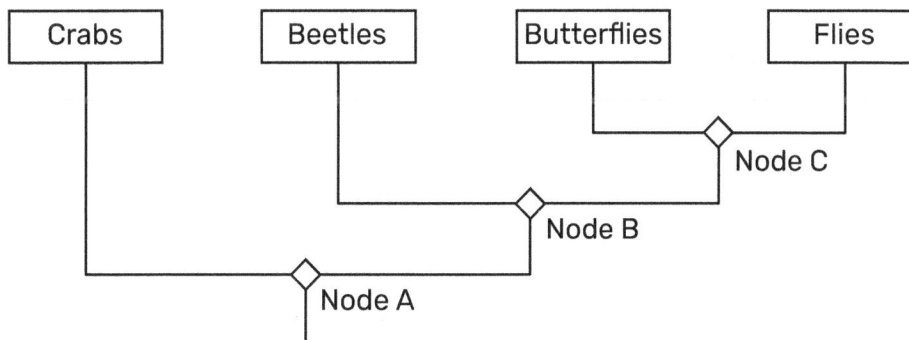

The nodes represent the common ancestors of all the taxa "above" the node. The more recently two taxa share a common ancestor, the more recently their lineages split, and therefore the more closely related the taxa are.

How do you know how recently two taxa shared a common ancestor? By looking for patterns in the traits they share. When a trait evolves in a population, it is passed down to all of the lineages that descend from that population. In this way, different groups of taxa come to share traits that other taxa—that descended from a different common ancestor—lack. When a trait is shared by a very small group of taxa, it usually means that those taxa are very closely related.

Look at figure 8.2. When you read a tree from root to tip, if you come across a trait that means that all of the taxa "above" that trait share it. For example, in the tree below, dogs, cats, and chipmunks share the group trait "has fur." Fur evolved in the common ancestor population of all mammals, so all mammals share it.

Figure 8.2. Tree of Animals with Fur

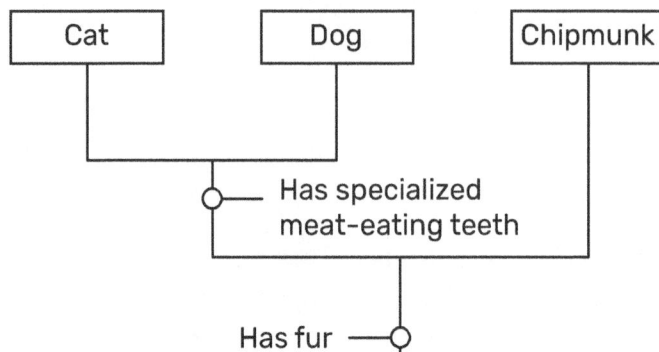

Dogs and cats share the group trait "specialized meat-eating teeth." This trait evolved in a common ancestor of dogs and cats, which lived more recently than the common ancestor of all mammals. Other mammals do not have this trait. So cats and dogs belong to two groups: the mammal group because they have fur and the dog-cat group because they have specialized meat-eating teeth. From this evidence, we can infer that cats and dogs are more closely related to each other than to mammal taxa that don't have specialized meat-eating teeth, like chipmunks.

2. Use the information in table 8.2 to determine which animal groups possess the trait at the top of each column in table 8.3. The first group trait ("has specialized cells with nuclei") has been completed for you. Once you have filled in all of table 8.3, answer the questions.

Table 8.3. Animal Groups that Share the Same Characteristic(s)

Trait	Has special-ized cells with nuclei	Has tissue	Has organs	Has a head	Has jointed arms and legs
In the column under each trait, list all the taxa that have that trait. The first trait is completed as an example.	Beetles Crabs Jellyfish Snails Sponges Seastars				

Q18. Based on your completed table 8.3, which two taxa are the most closely related?
Hint: Which pair of taxa share the most traits that other organisms don't have?

Q19. Based on your completed table 8.3, which two taxa are the most distantly related?
Hint: Which pair of taxa share the FEWEST traits that other organisms don't have?

3. Finally, take the information from table 8.3 and use it to build a phylogeny for these organisms in figure 8.3. The first two boxes have been filled in for you. These are the two groups of species that have the most traits in common. The next box to fill in would be the one next to "crabs" on the left. Write the name of the type of organism with the next most characteristics in common with the crabs and beetles. Keep the process going until you have filled out all the boxes with a name of a group of organisms in table 8.3. Then write the traits that each group would need to acquire to be different from the organisms before them in the box.

Figure 8.3. Build a Phylogeny for Organisms

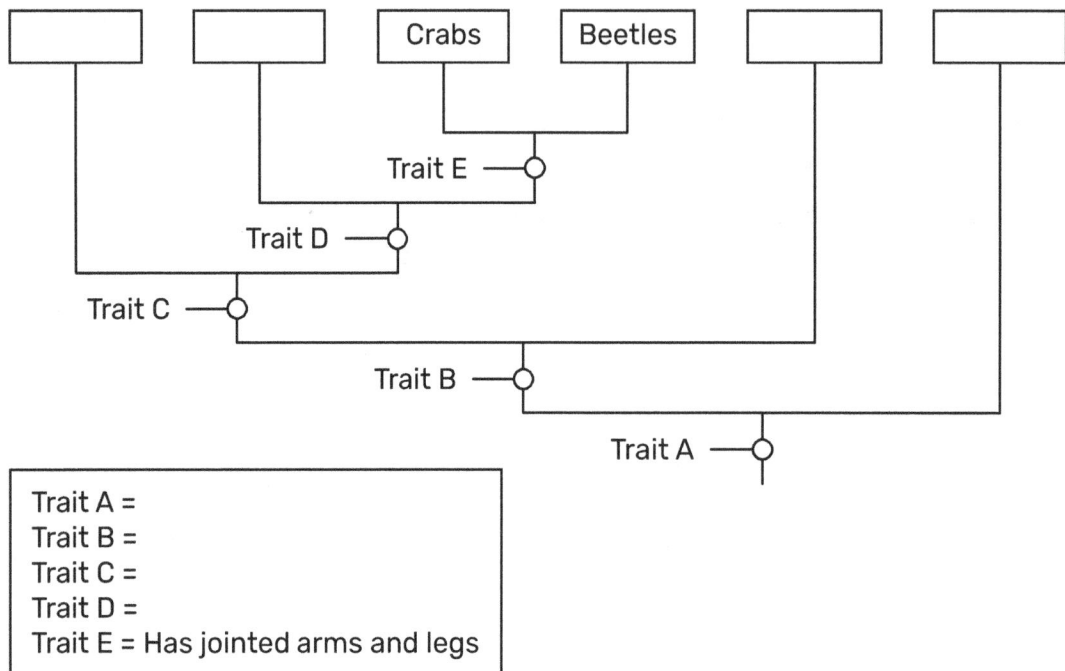

Trait A =
Trait B =
Trait C =
Trait D =
Trait E = Has jointed arms and legs

Exercise 8.2. Building a Phylogeny

The evolutionary history of animals is an exciting series of events that both occurred over millions of years and all at once. That is, even though evolutionary events are often presented as a timeline, first came species A then came species B, then came species C, in actuality, at the same time species C was diverging from a common ancestor with species B, so too was species A undergoing different evolutionary pressures causing it to diverge into species D and E. And so the evolutionary history of life on earth should be considered to be more like a giant bush (or tree) than a timeline, as seen in figure 8.4.

Figure 8.4. Evolutionary Tree of a Species

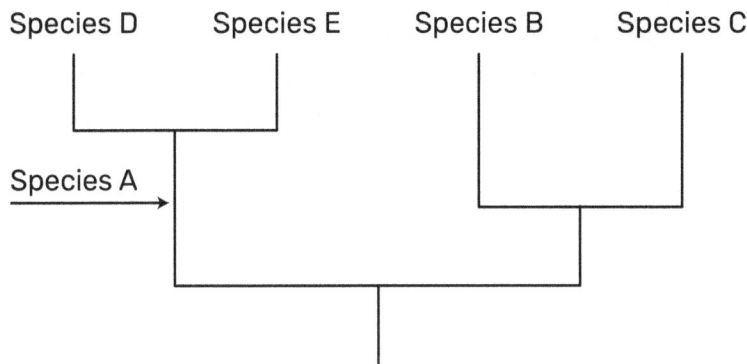

A. Building a Phylogeny from Your Observations

The lab tables are loosely organized according to habitat. You'll see representative samples of the diversity of animal life on planet Earth. One table represents a terrestrial environment, one is a marine environment, and one shows a freshwater habitat.

Your objective for this exercise is to become acquainted with the variety of animal life and some of the significant characteristics which help us to understand and organize them. Review your pre-lab here as you consider what you might look for. Information cards have been placed with a few of the samples with the scientific name indicated on the top and various facts on the bottom. Please observe and think about the specimen before turning the card over.

Guidelines Before We Begin

❑ This lab is an adventure and an exploration. This means that you may want to touch some of the samples. This is great, except where specifically stated otherwise. In general, we want to be respectful of the material and keep it in good shape. There are some live animals which can also be handled, but please ask for instructor assistance. Finally, we will use the buddy system, so you will team up with one other student as you venture out into this wilderness.

❑ Find types of animals that exhibit the traits listed below from Procedure Steps 1–6 (symmetry, body coverings, methods of locomotion, systems to support the body, different feeding structures, and reproductive strategies).

» For each of these traits there are three or four different ways that trait can be exhibited, or 19 different traits listed total in the boxes below. For instance, the trait symmetry could be (1) asymmetrical, (2) radially symmetrical, or (3) bilaterally symmetrical.

❑ Your task is to find an example of an animal from the collection before you that displays that trait. Sketch the animal and write its name in the box.

» See if you can use 19 *different* animals to fill the boxes below. For example, rather than using the turtle for "Body covering," "methods of locomotion," and "eating structure," use the turtle for "body covering," the bat for "methods of locomotion" and the sea star for "eating structure." You should not need to look up information about these organisms from the internet, rather use what you already know, what you learned from lecture and your textbook, and your powers of deduction, to determine which organisms have which traits. The process of observation and discovery is more important to this exercise than being "right."

Procedure

1. Find three examples of different **symmetry** and then sketch and label them in table 8.4.

Table 8.4. Examples of Symmetry

Asymmetry	Bilateral symmetry	Radial symmetry

a. Now look around for other organisms which share the same symmetries as those you listed above. List their names near your sketch in table 8.4.

Q20. What type of symmetry do most animals displayed here exhibit? What do you think might be an advantage to this type of symmetry?

2. Find four examples of different **body coverings**. In table 8.5, sketch an animal with that type of body covering (one per box) and label the box with the type of body covering and name of the animal.

Table 8.5. Examples of Body Coverings

Body covering 1	Body covering 2	Body covering 3	Body covering 4

a. Now look around for other organisms that share the same body coverings as those you listed above. List their names in table 8.5.

Q21. Name three roles these body coverings play in the lives of the animals and explain why each role is significant to the survival of the animal.

❑ _____

❑ _____

❑ _____

3. Find three examples of different methods of **locomotion** (what structures help the animals move?). Name each method in table 8.6 and then sketch an animal with that type.

Table 8.6. Examples of Locomotion

Method 1	Method 2	Method 3
_____	_____	_____

a. Now look around for other organisms which share the same means of locomotion as those you listed above. List their names in table 8.6.

Q22. How do the structures you describe above make the organism that has the structure better adapted to its environment?

8

4. Find three examples of different **systems to support and/or protect the body** of the organism. These are skeletal systems made from various structures—not just internal skeletons, like yours. Name each one in table 8.7 and then sketch an animal with that type.

Table 8.7. Examples of Support Systems

Support system 1	Support system 2	Support system 3
_____	_____	_____

a. Now look around for other organisms which share the same skeletal systems as those you listed above. List their names in table 8.7.

Q23. Compare these to your own skeletal system. How are they similar and how are they different?

8

5. Find three examples of different kinds of **feeding structures**. (Note: not every creature has a mouth and teeth, but all must "feed" in some way.) Name each type of feeding structure in table 8.8 and then sketch an animal with that type.

Table 8.8. Examples of Feeding Structures

Feeding structure 1	Feeding structure 2	Feeding structure 3
_____	_____	_____

a. Now look around for other organisms that share the same feeding structure as those you listed above. List their names in table 8.8.

Q24. Why is it advantageous for organisms to have their feeding structures close to most of the other sensory organs (e.g. your mouth is close to your nose and eyes and ears)?

Q25. In what situations is one type of feeding structure advantageous over another? Or in what type of situation is each feeding structure you examined well suited?

6. Find three examples of different **reproductive strategies**. In table 8.9, sketch the strategy or life cycle—**not** just the adult animal—and label them. Include the name and a description of the strategies you observed. A reproductive strategy would describe how or where fertilization, development, and birth take place—is fertilization internal or external? Is development internal or external? What type of egg does the species have? How does "birth" happen?

Table 8.9. Examples of Feeding Structures

Reproductive strategy 1	Reproductive strategy 2	Reproductive strategy 3

a. Now look around for other organisms that share the same feeding structure as those you listed above. List their names in the box above.

Q26. The type of egg an organism has corresponds to its reproductive strategy and the nourishment of its offspring. Describe the type of egg used in each of the reproductive strategies you found above.

❑ _____

❑ _____

❑ _____

B. Build Your Own Phylogeny

Having observed a variety of creatures, now let's go back and consider the phylogenetic relationship of these organisms. Which creatures have been present on earth for the longest period of time? Which have diverged from a common ancestor more recently?

Procedure

1. Pick ten of the types of animals you observed above.
2. Use table 8.10 as a guide to determine if the trait your organisms possess is a primitive trait, intermediate trait, or derived trait.

Table 8.10. Key to Primitive, Intermediate, or Derived Characteristics of Each Type of Trait Observed to be Used in table 8.11

Type of trait	Primitive trait	Intermediate trait	Derived trait
Symmetry	Asymmetry	Radial symmetry	Bilateral symmetry
Body covering	None	Shell/carapace	Skin/fur/feathers
Locomotion	None/drift	Swim	Walk/fly
Support system	Hydrostatic skeleton	External skeleton	Internal skeleton
Feeding structure	Absorption	Filter feeding	Mouth/ingestion
Reproduction	External fertilization and external development	Internal fertilization and external development	Internal fertilization and internal development

3. Write "P," "I," or "D" in the box in table 8.5, according to the type of trait possessed.

Table 8.11. Classification of Primitive, Intermediate, or Derived Characteristics for Each Trait Observed in Each Organism. Note: Use only "P,""I," or "D" to fill in each box according to the key in table 8.10.

Organism	Symmetry type	Body covering	Method of locomotion	Support system	Feeding structure	Method of reproduction	# P's	# I's	# D's

4. Examine table 8.11 and determine which organism has the most primitive traits.
5. Write the name of the most primitive organism at the end of line furthest to the left on the phylogeny in figure 8.6.

Figure 8.5. Example of a Phylogeny Using The Number of Primitive and Derived Traits to Establish Which Organism Appears Earliest (and second, and third, etc) in Evolutionary History

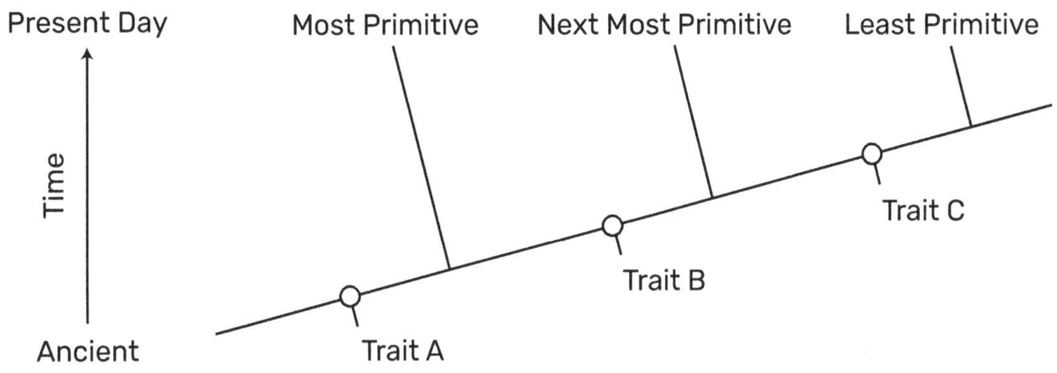

6. Which organism has the next most primitive traits? Write the name of this organism at the end of the second line to the left on the phylogeny in figure 8.6. Continue with this pattern until you have added all ten of your organisms from table 8.11 to the phylogeny.

a. The name of the organism with the largest number of primitive traits should be written at the end of the line furthest to the left and the name of the organism with the fewest primitive traits should be written at the end of the line furthest to the right. The names of the other organisms should be filled in according to the number of primitive and/or derived traits they exhibit.

Figure 8.6. Phylogeny for the 10 Organisms Assessed in table 8.5.

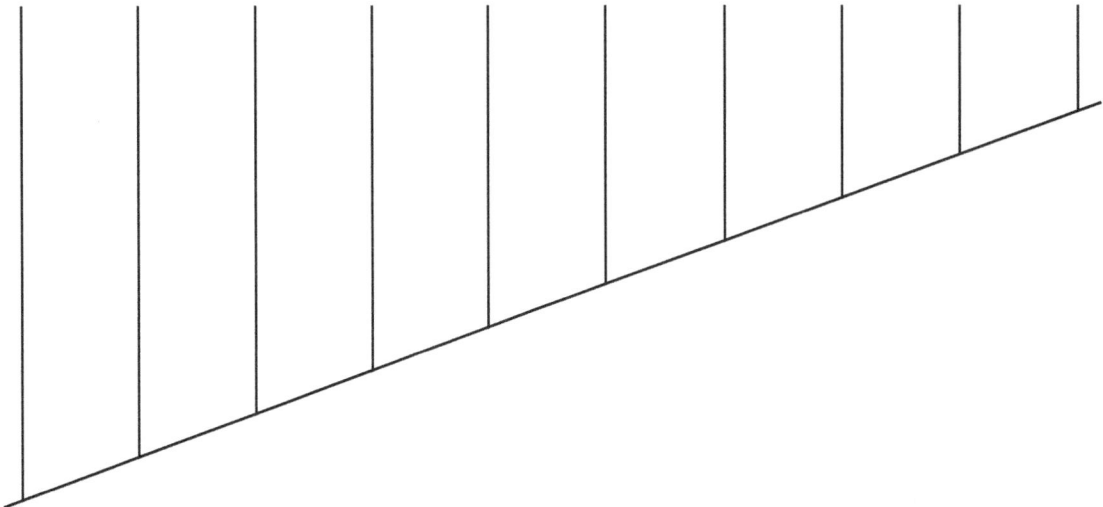

> This type of phylogeny conveys the same type of information depicted in the branching phylogeny in Exercise 8.2.A. The most primitive or "oldest" organisms are found at the end of the longest branch on the tree, and the most derived (or least primitive) organisms are found at the end of the shortest branches on the tree. The organisms that are more ancient are written at the end of the lines on the left and the organisms that diverged more recently are written at the end of the lines on the right.

Exercise 8.3. *Shape of Life* Videos

To understand the great diversity of animal life and to begin to understand the relatedness among animal phyla, we will watch a video series from the website *Shape of Life* that detail the set of characteristics and behaviors that distinguish one phylum from another. For this lab, you must watch the following five videos, which will take approximately 75 minutes. If lab time runs out before you watch these videos, your instructor may have you finish this exercise as homework.

The videos are available from the Phyla tab on the *Shape of Life* website. Scan the QR code or visit the URL listed to access the videos. You will need to click on each video to view it. As you watch, answer the questions relating to that video.

A. Sponges: Origins

Visit go.chemeketa.edu/sponges or scan QR code.

Q27. What happened 3.5 billion years ago? What sort of life dominated Earth for 2.5 billion years?

Q28. What was the first animal like? Describe the ancient sponge.

Q29. Describe the following innovations seen in sponges:

❑ Collagen

❑ Spicules

❑ Filter feeding

❑ Choanocyte

❑ Flagella

Q30. How do sponges eat? How do they reproduce?

B. Terrestrial Arthropods: The Conquerors

Visit go.chemeketa.edu/arthropods or scan QR code.

Q31. Breathing air was a major challenge in adapting to life on land. How do scorpions manage to breathe air? How do millipedes?

Q32. What did the first terrestrial arthropods eat?

Q33. What type of terrestrial arthropod was the first hunter?

Q34. Why do dragonflies go through metamorphosis?

Q35. What proportion of all animals do the flying insects comprise?

- ❏ _____
- ❏ _____
- ❏ _____

Q36. List two roles that terrestrial arthropods play in the environment.

C. Molluscs: The Survival Game

Visit go.chemeketa.edu/molluscs or scan QR code.

Q37. What is the purpose of each of the following in molluscs?

- ❑ Foot

- ❑ Radula

- ❑ Mantle

Q38. Describe the nautilus.

Q39. What have the squid gained by losing their shell? Why did they lose it?

Q40. What are chromatophores? What allows the octopus to use its chromatophores?

8

D. Echinoderms: The Ultimate Animal

Visit go.chemeketa.edu/echinoderms or scan QR code.

Q41. Describe the 5-point symmetry of the following organisms:

❑ Sea star

❑ Sea urchin

❑ Sea cucumber

Q42. Describe the skeleton of the sea star.

Q43. What is the ecological role of sea urchins?

Q44. How do sea stars eat?

E. Chordates: We're All Family

Visit go.chemeketa.edu/chordates or scan QR code.

Q45. What allowed the chordates to get so much bigger than animals in other phyla?

Q46. What is a notochord?

Q47. Who were the first chordates? What traits do humans share with these first chordates?

Q48. What other innovations do we see in vertebrates?

Q49. What advantage did fishes have over the organisms that came before them?

Q50. What vertebrate creature was first to walk on land? From what did it evolve?

Q51. How are reptiles able to reproduce away from water?

Q52. To what group of primates are humans most closely related? How are we similar?

Q53. What sets humans apart from other primates?

8 | Applying What You've Learned: Animal Diversity

1. As a population of a species evolves, can it maintain some traits in a primitive state while other traits it has are derived? Why or why not?

2. What traits do sponges have that place sponges at the base of the animal tree phylogeny? Why are these characteristics considered to be primitive traits?

3. Which organism that you observed in Exercise 2 had the largest number of derived traits? Which phylum does this organism belong to? You may need to look this up. What are the key characteristics of organisms in this phylum?

4. Convergent evolution is the process by which species that are not closely related evolve similar structures in response to similar environmental conditions or physiologic needs. A classic example of convergent evolution are the spines on plants in the family Cactaceae in southwestern US and the spines on plants in the family Euphorbiaceae in southern Africa. Look back through the species you saw today and explain one example of convergent evolution.

5. Of the six types of traits examined in Exercise 2 (symmetry, body covering, locomotion, support system, feeding structure, reproduction), which one was most helpful in developing your phylogeny? Why? Which trait was least helpful in developing your phylogeny? Why? Which trait was most confusing? Why?

A Appendix: Compound Microscope Introduction and Making a Wet Mount

At some point during the term, you will be required to use a microscope to complete the necessary analysis of lab materials. Your instructor may help you prepare for these labs by using this exercise that allows you to explore how microscopes work or by assigning a take-home lab that introduces these concepts through a virtual microscope. Whichever method is chosen to help you learn the names of the parts of the microscope and how each part assists you in magnifying the object you wish to see, the most important aspect of this preparation is that it will allow you to explore the microscopic world. With the skill of proper microscope use in your personal toolkit, you will be able to identify and quantify single-celled microscopic life forms, as well as cells and tissues from multicellular organisms. The microscope is an essential tool for the biologist and this exercise will help you master its use.

By the end of this exercise, you should be able to:

- ❑ Describe the function and use of the major parts of a microscope.
- ❑ Use a microscope to investigate the structure of very small objects.
- ❑ Demonstrate the proper way to clean a lens, focus on an object and transport and store your microscope.

Using a Compound Microscope to See Protists

Most protists are very small, and many are made up of a single cell. To observe protists and the details of their cells, we must learn to use a microscope. Several different types of microscopes can be used to observe cells. In this class, we will use both compound microscopes and dissecting microscopes. A **dissecting microscope** provides three-dimensional details at rather limited magnifications using either light reflected from surfaces or light transmitted through transparent specimens. A **compound microscope** provides two-dimensional details at much higher magnifications by gathering light that has passed through a thin specimen.

Figure A.1. Diagram of a Compound Microscope

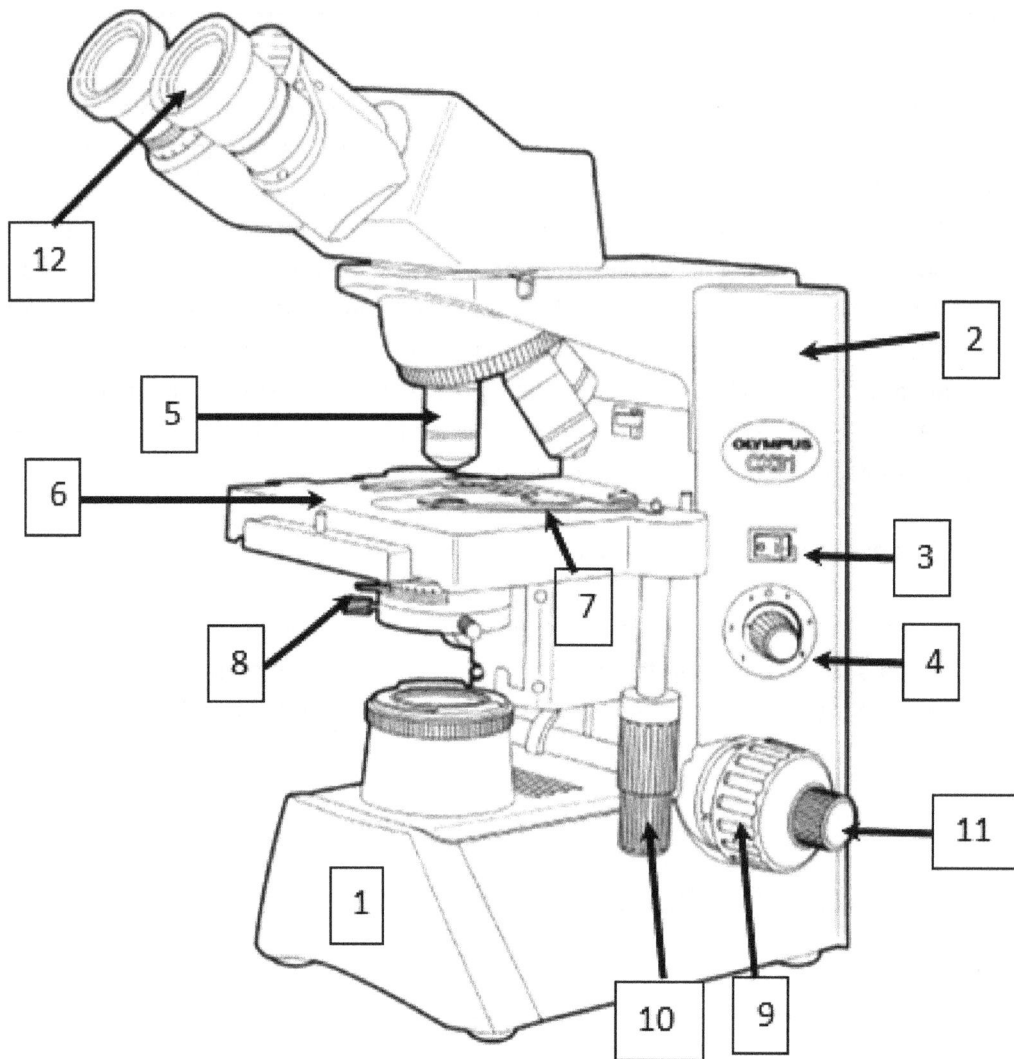

A. Microscope Parts

Numbers correspond to those in figure A.1.

1. **Base**: supports and bears the weight of the microscope. When carrying a microscope you should have one hand under the base while the other hand supports the arm of the microscope.
2. **Arm**: supports the body tube and is used to carry the microscope.
3. **On/Off switch**
4. **Rheostat**: controls the light intensity, but should usually be turned all the way on for bluest light and sharpest image.
5. **Objectives**: lenses at the bottom of the nosepiece. Scanning, Low Power, and High Power.

PLEASE NOTE: All observations should commence with scanning or low-power magnification.

6. **Stage**: supports the specimen over the hole that admits light from below.
7. **Mechanical stage feet**: holds the slide in place, and controls the movement of the slide.
8. **Iris diaphragm**: regulates the amount of light passing through the specimen.
9. **Coarse focus knob**: moves the nosepiece up and down above the specimen. It is the larger of the two knobs. Since the coarse adjustment moves the nosepiece a large distance, it is used only to focus with the scanning and low-power objective lens.
10. **Mechanical stage adjustment knobs**: move the slide on the stage.
11. **Fine adjustment**: moves the nosepiece only a small amount. It is the smaller knob. It is used to bring the specimen into exact focus. It is used only after the specimen is brought into view by coarse adjustment.
12. **Ocular**: contains lenses to magnify the specimen. Usually the ocular lens magnifies the specimen ten times (10x).

B. How to Use a Compound Microscope

1. Remove the microscope from the cabinet with two hands—one under the base (1) and one in the handle on the arm (2). Place the microscope on the counter, remove the dust cover, and store it in the cabinet.
2. Plug the microscope in and turn on the light (3). Set the rheostat (4) that controls the light brightness, turn it up to about "5" and leave it there.
3. Ensure that the scanning (smallest lens and lowest power) objective (5) is rotated into place over the hole in the stage. If not, rotate the nosepiece until you feel the objective click into place.

4. Place your slide on the stage (6) between the metal "feet" (7). The slide should go between (not under the feet) with the specimen on top and in the center of the hole in the stage.
5. Reduce the brightness by adjusting the diaphragm (8; the black plastic lever). Move the lever all the way to the right, so that it appears darkest when looking through the microscope, then slowly move the lever to the left until it just stops getting brighter. This should provide you the greatest balance of contrast and resolution while viewing slides.
6. Raise the stage using the coarse focus knob (9) until it is as close to the slide as possible. Ensure that the lens does not touch the slide.
7. Focus by rotating the coarse focus knob while looking through the eyepiece. You will lower the stage until you see a sharp image.
8. When the image is in focus, use the mechanical stage adjustment knobs (10) to move the slide so that the part you want to see is exactly in the middle of the lit area (field of view).
9. Use the fine focus knob (11) to sharpen the image.
10. If you want to see greater detail, rotate the nosepiece so that the low-power objective (10x) clicks into position. Adjust the focus with the fine focus knob. DO NOT LOWER THE STAGE BEFORE YOU SWITCH OBJECTIVES. If you focus it on scanning, it will not hit the slide on low or high power.
11. Go through the same process of focusing and centering before you go to high power (40x).

If you cannot get a clear, focused image with one of the objectives, clean off the lens with lens paper and lens cleaner (found in the drawers above your microscope cabinet). Never use any other paper, cloth, your finger, tap water, or saliva. To clean the lens, add a drop of lens cleaner to the tissue and then gently rub on the surface of the ocular lens (eyepiece), objective lens, condenser, or slide.

> If you have any difficulty with the microscope after going through these instructions, ask your instructor for help.

C. Putting Away Your Microscope

1. Rotate the nosepiece until the shortest (scanning) objective is in place and lower the stage to its lowest point.
2. Remove the slide from the stage and clean and dry it and put it back where you got it.
3. Clean off any water on the stage with a paper towel or absorbent paper.
4. Coil the cord on the back of the microscope, put the dust cover on and return it to the cabinet.

D. Letter "E" Exercise

1. Get a clean slide and clear plastic cover slip—wash and dry them if necessary.
2. Cut a small word out of a newspaper that includes the letter "e" (lowercase, standard print size).
3. Create a wet mount of your word.
   a. Put a drop of water on the slide with an eyedropper.
   b. Use forceps (tweezers) to place the word in the water and carefully lower the cover slip onto the drop trying to trap as little air as possible.
   c. If necessary, you can add more water by putting a drop on the slide right at the edge of the cover slip.
4. Hold the slide with the word upright and draw the letter "e" just as it appears to your unaided eye in its normal orientation.
5. Follow the directions in part B to observe the letter "e" with the scanning objective of your microscope.
6. In figure A.2, sketch the letter "e" as it appears when you view it with your eye and in the orientation that you see as you look through the microscope.

Figure A.2. Sketch of Letter "e"

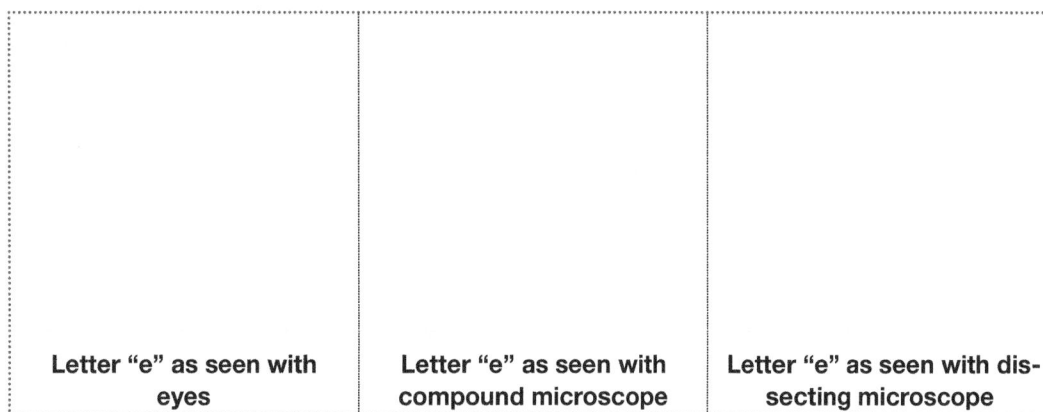

Letter "e" as seen with eyes	Letter "e" as seen with compound microscope	Letter "e" as seen with dissecting microscope

Q1. Compare your drawings. How does a compound microscope affect the image of the object you are viewing?

Q2. Move the slide away from you while you look at it through the microscope. Which way does the image move?

Q3. Move the slide to the left. Which way does the image move?

Q4. Why is it important to understand how the compound microscope affects the orientation of an image?

E. Making a Wet Mount with a Protist

1. Get a clean slide and clear plastic cover slip—wash and dry them if necessary.
2. Using the pipette provided for your protist sample, probe near the food source for the protist (usually a grain of wheat) and draw up about a cm worth of water into the pipette.
3. Squeeze the pipette over the center of your slide, creating a small circle of the sample about the size of a dime.
4. Gently lay the coverslip over the liquid, taking care not to create air bubbles.
    a. If necessary, you can add more water by putting a drop on the slide right at the edge of the cover slip.
    b. If you have too much water, you can touch a tissue to the edge of the cover slip and draw out some of the excess.
5. Follow the directions in Part B to place the slide properly on the stage and view the protists.
6. Return to the lab for specific questions relevant to this lab.

B Appendix: Key to Selected Woody Plants & Ferns in Western Oregon

This key is based largely on leaf characteristics and does not include all of the woody plants encountered in Western Oregon, but it should include the common ones found around the Willamette Valley and the Northern Oregon Cascades. Use the ruler on p. 184 if needed. *Low-growing herbaceous plants, generally those with flexible or fleshy green stems, are not included.* – Doug Ure (revised 7/14/2013)

* Naturalized non-native species.

Start Here!

1. a) Leaves are long, thin needles or short, overlapping scales — *go to 2*
    b) Leaves are flat and broad forming a blade with central vein(s) — *go to 17*

2. a) Leaves are needles — *go to 3*
    b) Leaves are scales — *go to 15*

Conifers

3. a) Two or more needles are bundled with a papery sheath around the base — *go to 4*
    b) Needles are attached to the branch individually (not gathered in bundles) — *go to 6*

4. a) Needles are two to a bundle — **Lodgepole Pine**
    b) More than two needles to a bundle — *go to 5*

5. a) Needles three to a bundle — **Ponderosa Pine**
    b) Needles more than three to a bundle — **Western White Pine**

6. a) Needles lie in <u>two distinct rows</u> on opposite sides of the branch forming a flat surface (Branch looks like it has been pressed with an iron.) — *go to 7*
    b) Needles point in all directions around the stem like bristles of a "bottle brush" or point upward like a "hair brush" or flare in two spreading layers from the two sides — *go to 9*

7. a) Needle tip has a small hook or point – **Western/Pacific Yew**
    b) Needle tip is rounded or has a small notch – *go to 8*

8. a) Needle is glossy green above with two white stripes below – **Grand Fir**
    b) Needle has white stripes on upper and lower surfaces – **White Fir**

9. a) Needles are sharply pointed and stiff (4-cornered in cross section) piercing your skin when you grab the branch — **Englemann Spruce (Cascades)** and **Sitka Spruce (Coast Range)**
    b) Needles are blunt and more or less flexible (not sharp and stiff) — *go to 10*

10. a) All needles along a section of stem (between two adjacent brahches) are almost the same length — *go to 11*

    b) Needles vary greatly in length between branching points of the stem (some as short as half the length of those right beside them) — *go to 14*

11. a) Needles are stiff and curve upward from lower surface sticking up on the upper surface like bristles of a hairbrush — *go to 12*

    b) Needles are flexible and stick out all around the branch forming a "bottle brush" appearance, which may be slightly flattened top and bottom – *go to 13*

12. a) Needles have a groove on the upper surface and two white lines on the lower surface, base of needle is sharply curved like a hockey stick. Found at higher elevations — **Noble Fir**

    b) Needles have two white lines on the upper and on the lower surface, needle is broader at the tip giving it a club-shaped appearance and base is only slightly curved. Found at high elevations — **Subalpine Fir**

13. a) Needles are tapered toward the tip, have a bluish cast, point from branch in all directions. Typically found growing with subalpine fir at high elevations — **Mountain Hemlock**

    b) Needles are narrowly linear, light to dark green, point all around the branch or toward the sides in spreading cross-sectional wedge. Bark of mature trees has deep, orangish, vertical cracks — **Douglas Fir**

14. a) Needles are glossy, dark green in color. Longer needles spread to sides from branch forming a flat lower surface. Shorter needles cluster along the upper center of the branch and point toward the tip. Found at higher elevations — **Pacific Silver Fir**

    b) Needles are flattened, thicker in the middle with a rounded tip, less than ¾-inch long. Dull green and roughly point toward the sides of the branch. Found growing at low to middle elevations — **Western Hemlock**

15. a) Overlapping scales wrap around branchlets forming branching green cylinders. Seeds are produced in "berries." Tree of dry sites east of the crest of the Cascades – **Western Juniper**

    b) Overlapping scales form flattened branchlets. Seeds are produced in small, brown cones on branch tips – *go to 16*

16. a) Pairs of scales on the branch twice as long as wide forming an elongated "Y". Flattened branchlets extend from mature branch at irregular angles. Drier sites of high Cascades and eastern slope — **Incense Cedar**

    b) Scales are about the same length and width. Flattened branchlets droop from mature branches forming a sloping, regular "shingled" appearance to the branches. Moister sites on western slope of the Cascades and entire Coast Range — **Western Red Cedar**

17. a) Leaves (fronds) are each a single stalk (rachis) arising from an underground stem and divided into many (more than 20) small blades (leaflets) — *Ferns, go to 18*

    b) Leaves are each attached to an above-ground stem and consist of a single blade or fewer than 20 small blades (leaflets) – *Broadleaf Trees and Shrubs, go to 24*

Ferns

18. a) Central stalk branches in a palmate pattern forming a fan of small, flat leaflets. Stalk is dark brown/black — **Maidenhair fern**
    b) Central stalk is unbranched or branches pinnately and is some other color — *go to 19*

19. a) Fronds (leaves consisting of a long stalk with multiple leaflets attached—think of a palm frond) in a whorl (radiating in a wagon-wheel spoke arrangement viewed from above) from a central (often underground) stem — *go to 20*
    b) Fronds arise individually from the ground (from a branching underground or hidden stem), no radial/whorl arrangement— *go to 22*

20. a) Frond consists of leaflets with fine marginal teeth. Leaflets in two rows along a central stem. Each leaflet is attached to the central stem by a short stalk, but the base of each leaflet is not attached except by a small stalk (free) — **Sword Fern**
    b) Frond is not like in "a" — *go to 21*

21. a) Frond consists of simple blade-like leaflets in two rows, each leaflet with a smooth edge and attached to the central stem by a broad base — **Deer Fern**
    b) Frond consists of two rows of deeply divided and branched leaflets that are shortest at the base and tip of the frond and longest in the middle — **Lady Fern**

22. a) Frond consists of two rows of simple, tapered, blade-like leaflets attached to the central stalk by a broad base (adjacent blades blend into one another) — **Licorice Fern**
    b) Frond not like this — *go to 23*

23. a) Single stalk divides pinnately into bare lateral branches which form attachment for blades starting several inches from the central stalk — **Bracken Fern**
    b) Stalk divides pinnately into secondary stalks which form attachment for blades immediately adjacent to the central stalk — **Spiny Wood Fern**

Broadleaf Trees and Shrubs

24. a) Thorns or spines on the stem (may be only a few) — *go to 25*
    b) Stem lacking thorns or sharp spines — *go to 30*

25. a) Leaves consist of a single blade attached to a stalk (petiole), thorns are stiff, long and woody – *go to 59*
    b) Leaves are compound (multiple leaflets attached to one stalk/petiole) — *go to 26*

26. a) Most leaves consist of three leaflets, or if more have pairs of leaflets, attached along one central petiole (pinnate) — *go to 27*
    b) Most leaves consist of five or more leaflets radiating out from a single attachment point at the end of the petiole (palmate). Note: some leaflets may have short secondary stalks attaching them to the petiole — *go to 29*

27. a) Three leaflets per leaf — *go to 28*
    b) More than three leaflets per leaf — **Wild Rose (several species)**

28. a) Stem is upright with thin, peeling, orangish-brown bark and scatter thorns on older, mature stalks – **Salmonberry**

    b) Stem is thin and flexible, green, trailing across the ground and other plants — **Trailing Blackberry (several species)**

29. a) Leaflets are ovoid (oval-like) with fine teeth along the margin — **Himalayan Blackberry***

    b) Leaflets are deeply notched, looking almost lacy — **Evergreen Blackberry***

30. a) Leaves have sharp spines along the edges (actually feel like <u>needles that will pierce your skin</u>) — *go to 31*

    b) Leaves lack sharp spines along the edges (a sawtoothed margin is not spines) — *go to 32*

31. a) Each leaf is simple (consisting of one blade) and leaves alternate along the branch – **Holly***

    b) Each leaf is compound (consisting of several <u>pairs of opposite leaflets</u> and a terminal leaflet). Compound leaves are attached to the stem alternately – *go to 71*

32. a) All or some leaves are compound (consisting of several leaflets attached to a single petiole) — *go to 33*

    b) Leaves are simple (consisting of a single blade attached to a petiole) — *go to 36*

33. a) Each leaf consists of three leaflets (with rounded, slightly lobed margins) — **Poison Oak**

    b) Not as in "a," each leaf consists of more than three leaflets — *go to 34*

34. a) Blades have a <u>sawtoothed margin</u> –*go to 72*

    b) Blades have a <u>smooth margin</u> (no teeth) — *go to 35*

35. a) Leaflet blades are over one inch long, all leaves are compound, young stems are grayish and lack longitudinal ridges — **Oregon Ash**

    b) Leaflet blades under one inch long, some are simple, young stem is green with longitudinal ridges — **Scotch Broom***

36. a) Venation (major vein pattern) is palmate (several veins radiate out from the petiole at the base of the blade) — *go to 37*

    b) Venation is pinnate (there is a single major vein up the middle of the blade and secondary veins branch off of it) — *go to 43*

37. a) Leaves are dark to medium green and glossy on the upper surface. Blade has an angular outline or may have deep notches between angular lobes. <u>Plant grows as a vine across the ground or climbing other vegetation</u> — **English Ivy**

    b) Leaves are light green. <u>Growth form is an upright tree or shrub, not a vine</u> — *go to 38.*

38. a) Leaves are oval with a central vein and a pair of secondary veins that curve toward the tip of the leaf (like a three-pronged pitchfork). Leaf margin is finely toothed, leaf surface is smooth, shiny and sticky with a spicy odor, bark is greenish/gray – **Snowbrush**

    b) Leaf and vein pattern not as above – *go to 39*

39. a) Leaves feel distinctly fuzzy – *go to 70*

    b) Leaves do not feel fuzzy — *go to 40*

40. a) Many mature leaves are large (over six inches across) with deep notches and slightly rounded lobes — **Bigleaf Maple**
    b) Largest leaves are under six inches across — *go to 41*

41. a) Leaves have blunt, rounded, lobed margin – *go to 42*
    b) Leaves have 7–9 sharp, pointed lobes – **Vine Maple**

42. a) Palmate leaf has five major veins– **several species of Currant**
    b) Palmate leaf with three major veins/lobes – **Pacific Ninebark**

43. a) Leaves are elongated and pointed and have yellow to golden fuzz on under surface (which may wear off as they age) — **Golden Chinquapin**
    b) Leaves may or may not be fuzzy but definitely lack the golden color on under surface — *go to 44*

44. a) Leaves are not toothed along the margins (edges) although they may be slightly wavy or lobed – *go to 45*
    b) Leaves are toothed along all or part of the margin (although the jagged edge may be very fine and difficult to see) – *go to 57*

45. a) Leaves are attached to the stem in pairs at each node (opposite) — *go to 46*
    b) Leaves are attached to the stem individually (alternate, one at each node) — *go to 48*

46. a) Leaf blades are small, generally less than 2 × 2.5 inches. Spindly shrub 1–2 m tall. – **Snowberry**
    b) Leaf blades are generally larger than 2 × 3 inches. Veins curve along the margins toward the tip of the leaf — *go to 47*

47. a) Small tree with gray bark, in the spring it has single, showy white "flowers" — **Pacific/ Western Dogwood**
    b) Many-branched shrub with red bark, clusters of small, white flowers in summer and small pinkish-white berries in the fall — **Red Osier/Creek Dogwood**

48. a) Leaves have two or more deep, rounded notches between major veins creating a multi-lobed shape – *go to 49*
    b) Leaves lack multi-lobed shape — *go to 50*

49. a) Leaf blades are tiny, less than one inch long. Shrub of the dry, Eastern slope and high desert – **Bitterbrush**
    b) Leaf blades are large, over two inches long. Deciduous tree of the Willamette Valley – **Oregon White Oak**

50. a) Largest leaf blades are over four inches long — *go to 51*
    b) Most leaf blades are clearly less than four inches long — *go to 55*

51. a) Leaves are dark green and leathery – *go to 69*
    b) Leaves are not dark green and leathery — *go to 53*

52. (Nothing should take you here and you go nowhere from here.)

53. a) Leaves are thin and strap-like, blades less than 1.5 inches wide and over four inches long tapering to a terminal point — **some species of Willow**

b) Leaves wider than 1.5 inches — *go to 54*

54. a) Leaves have <u>thick, distinct, straight, secondary veins</u> running parallel to one another toward the leaf margin (May have very fine teeth.) Leaf blade is oval, widest in the center. Upper surface has parallel grooves like a "washboard." Shrub or small tree — **Cascara "Buckthorn"**

b) Secondary veins are indistinct, not as in "a". Leaf blade tends to be broadest ⅔ of the way toward the tip and tapers narrowly toward the petiole. Shrub, up to 3 m tall. — **Indian Plum**

55. a) Oval leaves are dull green, thin (fragile) and blades are 0.5–1.5 inches long. Young stem is green with longitudinal ridges — **Red Huckleberry**

b) Most leaf blades are over two inches long — *go to 56*

56. a) Oval leaves are thick with indistinct veins. It is difficult to distinguish between upper and lower surface of the leaf. Old branches with reddish-brown peeling bark – **Manzanita**

b) Upper and lower surfaces of leaf are clearly different, with distinct veins. Old branches grayish, green or yellow — **Willows (several species)**

57. a) Oval leaves with a toothed margin and <u>one or two dark, swollen bumps on the petiole at the base of the blade</u> — **Bitter Cherry (and domestic cherries*)**

b) All leaves lacking bumps (glands) on the petiole, blade either lacks teeth or has them — *go to 58*

58. a) Mature leaf blades are tough and leathery, resisting folding, sometimes shiny on the upper surface – *go to 67*

b) Mature leaf blades are thin, relatively fragile (not leathery), and not shiny above — *go to 60*

59. a) Leaves are pinnate. Scattered woody thorns at the base of some leaf petioles – **Hawthorn**

b) Leaves are palmate – *go to 74*

60. a) Leaf is tear-drop shaped, similar to a "spade" on a playing card, broad at the base and tapering to a fine point at the tip — **Black Cottonwood**

b) Leaf is roughly oval, broadest near the middle of the blade — *go to 61*

61. a) Leaves are somewhat triangular with rounded teeth on the margin, almost like small lobes, one tooth at the end of each major vein. Plant is a shrub with very thin, arching stems — **Oceanspray**

b) Leaves and stems not as in "a". Leaves not roughly triangular — *go to 62*

62. a) Leaf is fuzzy and nearly round with a margin consisting of large teeth at the end of each major vain and smaller teeth forming the margin of each large tooth. Plant grows as a multi-stemmed, rigid shrub – **Hazelnut**

b) Leaf is ovoid, not fuzzy and has marginal teeth of only one size – *go to 63*

63. a) Leaf blade generally over four inches long and three inches wide. Grows as a single trunk tree. Bark tends to blotched in shades of white and grey due to growth of crustose lichens — **Red Alder**

b) Leaf blade usually less than four inches long and less than three inches wide – *go to 64*

64. a) Fine teeth only along the terminal ½ to ⅔ of the leaf blade (base lacks teeth) — *go to 65*

b) Fine teeth along the entire margin of the leaf blade— *go to 66*

65. a) Ovoid leaves are thin, same color above and below, loose, woody shrub to 5 m tall — **Saskatoon / Serviceberry**

b) Elongated leaves are dark above and gray/wooly below, dense, dense short shrub to 2 m tall. In summer with dense clusters of pink flowers on elongated vertical shoots. Dead flowers remain on brown stalks into winter — **Hardhack / Douglas Spirea**

66. a) Leaf blade is more than twice as long as it is wide — **Willow (several species)**

b) Leaf blade is less than twice as long as it is wide — **Pacific Crab Apple**

67. a) Mature large leaf blade is less than 1½ inches long – *go to 68*

b) Mature leaf blade is greater than 1½ inches long. Plant is and upright shrub. Dark green 2–4 inch-long oval leaves have a short petiole and are attached individually to the stem (alternate attachment) – **Salal**

68. a) Leaves attached to stem singly (alternate arrangement). Upright, densely branching shrub with slight reddish cast to stem. – **Evergreen Huckleberry**

b) Leaves attached to stem in pairs (opposite arrangement). – *go to 73*

69. a) Bark is orange and peeling on older branches, with green showing through beneath the thin orange layer on young stems – **Madrone**

b) Bark on older stems is grayish brown and lacking on younger stems so that they are simply green – **Pacific Rhodedendron**

70. a) Lobes of the leaf are pointed. – **Thimbleberry**

b) Lobes of the leaf are rounded and toothed – **Red Flowering Currant**

71. a) Leaves are <u>glossy green above</u>, usually consisting of consists of 5–9 leaflets. Plant is generally over 1 m tall – **Tall Oregon Grape**

b) Leaves are dull green above, usually consisting of nine or more leaflets. Plant is generally 0.5 m tall or less – **Creeping / Long-leaf Oregon Grape (Oregon State Flower)**

72. a) Pinately compound leaves are attached to the stem in pairs, two petioles per node (opposite leaf arrangement). Often a multi-stemmed shrub up to 7 m tall – **Elderberry**

b) Pinately compound leaves are attached to the stem one per node (alternate attachment). Multi-stemmed shrub up to 4 m tall – **Sitka Mountain Ash**

73. a) Plant is a trailing vine along the ground and trailing over downed logs. Shiny green one-inch roundish leaves are attached to the stem in pairs (opposite attachment) – **Twinflower**

b) Plant is a low-growing stiff shrub with grayish-brown cast to bark and dark green, stiff, oval leaves attached in pairs (opposite) – **Oregon Boxwood**

74. a) Largest leaves are 5 inches wide. Branching shrub up to 6 feet tall. Needle-like spines along the stem with longer ones at the nodes. – **Black Gooseberry**

b) Largest leaves up to 14 inches wide. Largely unbranched, thick, crooked stem with numerous spines up to ½ inch long. – **Devil's Club**

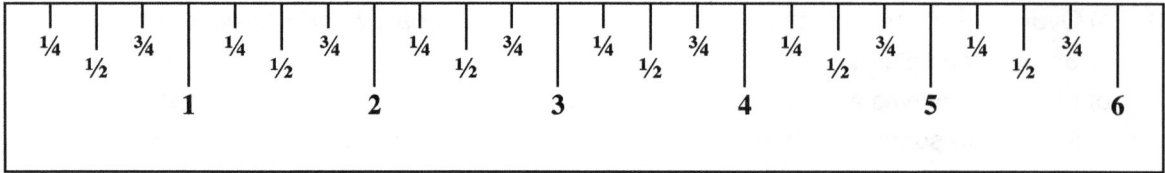

References

BBC Four. "Han Rosling's 200 Countries, 200 Years, 4 Minutes—The Joy of Stats." YouTube, November 16, 2010. https://www.youtube.com/watch?v=jbkSRLYSojo

BioInteractive. "Animated Life: Seeing the Invisible." October 21, 2014. https://www.biointeractive.org/classroom-resources/animated-life-seeing-invisible

BiteSizeBio. "History of Cell Biology." Published November 5, 2007. https://bitesizebio.com/166/history-of-cell-biology/

Christopher J. Burrell, ... Frederick A. Murphy, in Fenner and White's Medical Virology (Fifth Edition), 2017. https://www.sciencedirect.com/topics/medicine-and-dentistry/history-of-virology

Gapminder. "Don't Panic—The Facts About Population." November 17, 2023. https://www.gapminder.org/videos/dont-panic-the-facts-about-population/

George Washington University. "History of the Protists." Accessed January 1, 2024. https://www2.gwu.edu/~darwin/BiSc151/Protista/protists.html

Global Footprint Network. "Ecological Footprint Calculator." Accessed January 1, 2024. https://www.footprintcalculator.org/home/en

Microscope Master.com. "History of the Microscope." Accessed January 1, 2024. https://www.microscopemaster.com/history-of-the-microscope.html

National Geographic. "Coral Reefs 101." YouTube, November 17, 2017. https://www.youtube.com/watch?v=ZiULxLLP32s

National Oceanic and Atmospheric Association Pacific Marine Environmental Laboratory. "A primer oh pH." Accessed January 1, 2024. https://www.pmel.noaa.gov/co2/story/A+primer+on+pH

Plagiarism. org. "What is Plagiarism?."Published May 18, 2017. https://www.plagiarism.org/article/what-is-plagiarism

PopulationPyramid.net. "Population Pyramids of the World from 1950 to 2100." Accessed January 1, 2023. "https://www.populationpyramid.net/#google_vignette

Population Reference Bureau. "Data Sheets." Accessed January 1, 2024. https://www.prb.org/collections/data-sheets/

Shape of Life. "The Story of the Animal Kingdom: Phyla." Accessed January 1, 2024. https://www.shapeoflife.org/phyla

The Nature Conservancy. "Sprucing up the Appalachians." YouTube, July 29, 2024. https://www.youtube.com/watch?v=gEKpKUuczwE

www.ingramcontent.com/pod-product-compliance
Lightning Source LLC
Chambersburg PA
CBHW081516270226

40270CB00001B/1

* 9 7 8 1 9 5 5 4 9 9 6 3 7 *